William A. Tilden

Introducton to the Study of Chemical Philosophy

The Principles of Theoretical and Systematic Chemistry

William A. Tilden

Introducton to the Study of Chemical Philosophy
The Principles of Theoretical and Systematic Chemistry

ISBN/EAN: 9783337069360

Printed in Europe, USA, Canada, Australia, Japan

Cover: Foto ©Thomas Meinert / pixelio.de

More available books at **www.hansebooks.com**

TEXT-BOOKS OF SCIENCE

ADAPTED FOR THE USE OF

ARTISANS AND STUDENTS IN PUBLIC AND SCIENCE SCHOOLS

CHEMICAL PHILOSOPHY

New Edition of Watts' Dictionary of Chemistry.

WATTS' DICTIONARY OF CHEMISTRY.

Revised and entirely Re-written by H. FORSTER MORLEY, M.A. D.Sc. Fellow of, and lately Assistant-Professor of Chemistry in, University College, London; and M. M. PATTISON MUIR, M.A. F.R.S.E. Fellow, and Prælector in Chemistry, of Gonville and Caius College, Cambridge. Assisted by Eminent Contributors. To be published in 4 vols. 8vo. Vols. I. and II. (*A—Indigo*) now ready, price 42*s*. each.

THORPE'S DICTIONARY OF APPLIED CHEMISTRY.

By T. E. THORPE, B.Sc. (Vict.) Ph.D. F.R.S. Treas. C.S. Professor of Chemistry in the Normal School of Science and Royal School of Mines, South Kensington. Assisted by Eminent Contributors. 3 vols. price 42*s*. each. Vol. I. (*A—Dysodil*) now ready.

MODERN THEORIES OF CHEMISTRY.

By Professor LOTHAR MEYER. Translated, from the Fifth Edition of the German, by P. PHILLIPS BEDSON, D.Sc. (Lond.) B.Sc. (Vict.) F.C.S. Professor of Chemistry, Durham College of Science; and W. CARLETON WILLIAMS, B.Sc. (Vict.) F.C.S. Professor of Chemistry, Firth College, Sheffield. 8vo. 18*s*.

'This is, perhaps, the most important and profound philosophical treatise on one of the most fascinating and widely studied sciences... There could scarcely be a more acceptable book published than this translation, which commends itself by the manifest care exercised in its production.'—SCOTSMAN.

London: LONGMANS, GREEN, & CO.

INTRODUCTION TO THE STUDY
OF
CHEMICAL PHILOSOPHY

THE PRINCIPLES
OF
THEORETICAL AND SYSTEMATIC CHEMISTRY

BY

WILLIAM A. TILDEN, D.Sc.Lond., F.R.S.

PROFESSOR OF CHEMISTRY IN THE MASON COLLEGE, BIRMINGHAM
LATE LECTURER ON CHEMISTRY IN CLIFTON COLLEGE

NEW EDITION

LONDON
LONGMANS, GREEN, AND CO.
AND NEW YORK: 15 EAST 16th STREET
1890

All rights reserved

PRINTED BY
SPOTTISWOODE AND CO., NEW-STREET SQUARE
LONDON

PREFACE

TO

THE SIXTH EDITION.

TWELVE years have elapsed since the production of this little book. In the course of those years many facts have been added to our knowledge of chemical laws and phenomena, and many changes have been introduced into chemical theory. Thus all the gases have been liquefied, several new elements have been discovered, and the 'periodic law' has been accepted by the majority of chemists. Hence it has become necessary to subject the book to very thorough revision. This I have done, but at the same time I have not felt it desirable to alter its character or to increase its dimensions so much as to render it less suitable than before for those persons for whom it was originally designed, namely, for pupils in the advanced classes in schools and colleges, and generally for those students who have passed through a sufficient preparatory course of experimental instruction.

For the student who intends to pursue chemistry specially it may serve as an introduction to more comprehensive works, such as Professor Lothar Meyer's 'Modern Theories.'

In matters of theory it can hardly be expected that the

views expressed by one writer should agree entirely with those of another. I have, however, done my best to treat in an impartial spirit those parts of the subject upon which opinion is divided. I have also admitted very little matter of a speculative kind, and the single chapter which might be open to such an objection (Chap. XXII.) may be omitted by young students. In a few cases I have felt justified in stepping a little in advance of the ordinary orthodox teaching—as, for example, in connection with valency. With regard to classification, I have preferred to retain the arrangement of the elements in the most obviously natural groups, as being easier for the student than a system based rigidly upon the periodic law, which, however, is discussed in a separate chapter.

In the direction of thermo-chemical investigation great progress has been made ; but much yet remains to be done in this department of the science before the full significance of the observed facts will become manifest. Partly on this ground, and partly because the subject could not be treated in an instructive manner without making very extensive additions to the book, I have preferred to omit, for the present, a description of the experimental methods and all but the most obvious conclusions.

The short sketch of the classification of carbon compounds is, of course, wholly inadequate if regarded as representing the range of this extensive department of chemistry. It is introduced merely for the benefit of the general reader who may not have time for the serious study of this complicated subject ; it may also serve as a short syllabus for the use of classes in schools.

W. A. T.

BIRMINGHAM : *May* 1888.

EXTRACT FROM

PREFACE TO THE FIRST EDITION.

This little volume is primarily intended for the use of students. It aims at presenting a synopsis—brief indeed, and probably imperfect—of the leading principles of chemistry in such a form as to give the subject a more decided educational direction than has been hitherto customary.

In consideration of its peculiar fitness for developing the powers of observation, of reasoning, and of memory, no branch of experimental science deserves more emphatic recognition at the hands of educators than chemistry. In order, however, that its advantages may be reaped to the full, I believe that the methods of teaching very generally prevalent in schools require to be considerably modified. I think teachers ought to realise the fact that chemistry, as a school subject, is not taught with a view to its practical applications to medicine, manufactures, or the arts, but because the study is calculated to quicken the faculties of observation, to strengthen the memory, and to engender a power and a habit of continuous thought, as well as to arouse new interests and open up new fields to the imagination. It is of little consequence, in this view, whether or not the facts acquired can be turned to practical account ; but it is of prime importance that the phenomena brought under their notice, and the manner in which those phenomena are presented, should be such as will compel the pupils to think.

Notwithstanding that the book does not profess to be a complete treatise on the subject, its contents will, I believe, be found sufficiently comprehensive to afford a tolerably general view of chemical theory as it exists at the present time. My desire has been to assist the student in attaining to broad and philosophic views of chemistry as a whole, and to accustom him to regard it as one out of many branches of physical science rather than as a mystery standing apart from other studies

CLIFTON : *May* 1876.

CONTENTS.

	PAGE
TABLE OF WEIGHTS, MEASURES, AND TEMPERATURES	2

SECTION I.

CHAPTER I.

THE CONSTITUTION OF MATTER.

Molecules—Atoms—Chemical Affinity—Special Province of Chemistry—Solids—Liquids—Gases—Intermediate States of Matter 3

CHAPTER II.

FUSION.

Melting and Dissolution—Relation of Melting-point to Molecular Weight—Melting of Mixtures—Eutectic Mixtures—Cryohydrates 12

CHAPTER III.

SOLUTION.

Solution of Solids—Selective Power of Solvents—Fusibility and Solubility—Relation of Solubility to Temperature—Thermal Effects of Dissolution—Dissolution generally attended by

Contraction—Vapour Pressure of Solutions—Solutions of Coloured Salts—Theories of Solution—Crystallisation of Solutions—Water of Crystallisation—Supersaturated Solutions—Solution of Gases—Coefficients of Absorption—Law of Henry and Dalton—Case of Atmospheric Air . . . 16

CHAPTER IV.

LIQUID DIFFUSION AND DIALYSIS.

Crystalloids and Colloids—Dialysis—Soluble Colloidal Ferric, Silicic, and Aluminic Hydrates, &c. 29

CHAPTER V.

EVAPORATION AND EBULLITION.

Conditions of Ebullition—Relation of Boiling-point to Molecular Weight—Homologous Series—Fractional Distillation—Liquefaction of Vapours and Gases—Critical Temperature of Gases under Pressure 33

CHAPTER VI.

DIFFUSION AND DIALYSIS OF GASES.

Law of Diffusion—Graham's Experiments—Explanation of Diffusion by Molecular Kinetic Theory—Passage of Gases through Colloid Septa—Occlusion of Gases by Metals—Absorption of Gases by Porous Solids 40

CHAPTER VII.

RELATION OF GASES TO TEMPERATURE AND PRESSURE.

Law of Boyle—Law of Gay-Lussac and Charles—Examples—Absolute Temperatures—Law of Avogadro . . . 46

CHAPTER VIII.

SPECTRA—EMISSION SPECTRA.

Spectra of Solids, Liquids, and Dense Vapours are usually Continuous—Line Spectra of Ignited Gas—Spectral Analysis—Discovery of Elements by Spectroscope—Absorption Spectra—Fraunhofer's Lines—Heating and Chemical Effects of Radiant Energy—Luminosity of Flame—Theories of Davy and Frankland 53

EXERCISES ON SECTION I. 67

SECTION II.

CHAPTER IX.

ELEMENTS AND COMPOUNDS.

The Term Element in Chemistry only Provisional—Elemental Character of Chlorine, Potassium, Oxygen, &c., established by Experiments—Distribution of Elements in Nature—Table of Known Elements, with Symbols and Atomic Weights—Formulæ and Molecular Weights 70

CHAPTER X.

LAWS OF CHEMICAL COMBINATION.

Definite, Multiple, and Reciprocal Proportions—Gay-Lussac's Law of Volumes—Composition expressed in Percentages of Elements—Dalton's Atomic Theory 79

CHAPTER XI.

EQUATIONS—CLASSIFICATION OF REACTIONS.

Calculations of Weight and Volume in Chemical Changes—Classification of Reactions—Combination of Entire Molecules—Decomposition of Molecules—Metameric Change—Single Displacement—Double Decomposition—Substitution Compounds 86

CHAPTER XII.

CHEMICAL COMPOUNDS DISTINGUISHED FROM MIXTURES.

PAGE

Properties of Compounds always Different from those of their Constituents—Chemical Compounds have a Definite Composition—Fractional Crystallisation of Solids—Fractional Distillation of Liquids—Fractional Solution and Diffusion of Gases—Examples—Atmospheric Air is a Mixture . . 92

CHAPTER XIII.

NOMENCLATURE.

Names of Elements—Of Binary Compounds—Of Acids and Salts —Of Carbon Compounds 99

EXERCISES ON SECTION II. 105

SECTION III.

CHAPTER XIV.

ATOMIC WEIGHTS.

Experimental Determination of Combining Proportions—Oxygen, Carbon, Chlorine, Potassium, Silver, considered as Examples — Determination of Atomic Weight by Reference to Density of Gas or Vapour—Law of Dulong and Petit—Law of Isomorphism Atomic Weights of Oxygen, Carbon, &c., determined by Reference to Constitution of their Compounds . 109

CHAPTER XV.

MOLECULAR WEIGHTS AND FORMULÆ.

Formula deduced from Percentage Composition—Vapour Density and Molecular Weight—Synthetical Formation of Molecules of Oxygen, Hydrogen, &c.—Nascent State—Molecular Formulæ of Molecules—Synthesis and Analysis of Compounds— Saturating Power of Acids and Bases—Substitution Compounds 122

CHAPTER XVI.

TYPES—VALENCY.

Rational and Typical Formulæ—Constitutional Formulæ—Formula of Sulphuric Acid for Example—Valency or Combining Capacity—Atomicity—Valency of a given Atom varies according to Nature of Elements with which it is combined—Valency a Function of Temperature—Table of Principal Elements—Examples of Constitutional and Graphic Formulæ—Law of Even Numbers—Artiads and Perissads—Compound Radicles—Table of Valency of Common Salt Radicles 136

EXERCISES ON SECTION III. 153

SECTION IV.

CHAPTER XVII.

CONDITIONS OF CHEMICAL CHANGE.

In order that Chemical Action must occur between Two Bodies they must be in contact with each other—Complete Union only occurs in Liquid or Solid State—Influence of Temperature—Of the Solvent or Medium—Influence of Preponderant Mass—Action of a Third Body—Catalytic Actions . . 161

CHAPTER XVIII.

DISSOCIATION.

Nitrogen Peroxide—Ammonium Salts—Calcium Carbonate—Efflorescent Salts—Dissociation in Solution—Abnormal Vapour Densities—Explanation of Dissociation—Dissociation compared with Evaporation—Table of Examples . . 168

CHAPTER XIX.

THERMAL AND ELECTRICAL EFFECTS OF CHEMICAL ACTION.

Heat of Neutralisation of Acids—Effect of Successive Molecules of Base added to Polybasic Acids—Chemically Free or Unsaturated Bodies contain a Store of Energy—Principle of Maximum Work—Heat of Formation of sundry Chlorides, Bromides, Iodides, Oxides—Chemical Compounds contain generally Less Energy than the Materials from which they are formed—Exceptions—Electricity produced by Chemical Action—Electrolysis—Difficulties presented by some Phenomena of Electrolysis 176

CHAPTER XX.

COMBUSTION.

Modern Explanation of Combustion—Theory of Phlogiston—Igniting Points—Temperature of Flame—Heat of Combustion—Problems 186

CHAPTER XXI.

ISOMERISM.

Physical Isomerides—Polymerism—Metamerism—Theory of Asymmetric Carbon—Allotropic Forms of the Elements—Energy of Isomerides 192

CHAPTER XXII.

THEORIES REGARDING THE NATURE OF CHEMICAL AFFINITY.

General Characteristics of Chemical 'Affinity'—Three Chief Hypotheses: (1) Chemical Combination the Result of a Special Attraction; (2) Electro-Chemical Theories; (3) Atomic Motion—Chemical 'Affinity' compared with Adhesion—Molecular Compounds — Molecular Volume — Relative

Contents.

Strength of Affinity between Different Elements—The Chemical and Dynamical Equivalent distinguished—Avidity of Acids 203

EXERCISES ON SECTION IV. 215

SECTION V.

CHAPTER XXIII.

CLASSIFICATION OF ELEMENTS.

Division of Elements in Three Classes: Non-metals, Metalloids, Metals—Characters of Non-metals—Class 1. The Halogens—Class 2. Oxygen, Sulphur, Selenion—Class 3. Boron—Class 4. Carbon, Silicon—Class 5. Nitrogen, Phosphorus 217

CHAPTER XXIV.

DIVISION II.—METALLOIDS.

1. Hydrogen—2. Tellurium—3. Germanium, Tin, Titanium, Zirconium—4. Vanadium, Arsenic, Antimony, Bismuth—5. Niobium, Tantalum—6. Molybdenum, Tungsten, Uranium 244

CHAPTER XXV.

DIVISION III.—METALS.

General Characters of Metals—Class 1. Metals of the Alkalies—Class 2. Metals of the Alkaline Earths—Class 3. Zinc Group—Class 4. Silver, Mercury—Class 5. Thallium, Lead—Class 6. Aluminium Group—Class 7. Iron-Copper Group—Class 8. Platinum Group 256

CHAPTER XXVI.

THE PERIODIC LAW.

Prout's Hypothesis—Short Series of Related Elements—Dumas compares such Series to Homologous Series—Newlands—Lothar Meyer's Curve, showing Periodicity of Atomic Volume—Mendelejeff's Classification 284

EXERCISES ON SECTION V. 293

SECTION VI

CHAPTER XXVII.

CLASSIFICATION OF COMPOUNDS.

Acids—Determination of Basicity of Acids—Bases—Ammoniacal Bases—Salts 295

CHAPTER XXVIII.

DERIVATIVES OF AMMONIA.

Ammonium Theory—Amines—Constitution of Ammonium Compounds—Phosphines, Arsines, Stibines—Amides . . 307

CHAPTER XXIX.

CARBON COMPOUNDS.

Homologous Series—Hydrocarbons and their Haloid Derivatives—Alcohols—Primary, Secondary, Tertiary Alcohols—Phenols—Ethers—Aldehyds and Ketones—Acids—Basic Derivatives of Ammonia—Compound Ethers, or Ethereal Salts—Organo-metallic Compounds 314

EXERCISES ON SECTION VI. 334

INDEX 339

CHEMICAL PHILOSOPHY.

METRIC WEIGHTS AND MEASURES.

Prefixes.—Myria. 10,000
Kilo- 1,000
Hecto- 100
Deka- 10
( ) 1
Deci- . . . . $\frac{1}{10}$
Centi- . . . . $\frac{1}{100}$
Milli- . . . . $\frac{1}{1000}$

Inserting after the hyphen and in the brackets —
(1.) *Metre*, gives the measures of length.
(2.) *Litre*, ,, ,, capacity.
(3.) *Gram*, ,, ,, weight.

LENGTH.

1 Metre = 10 decimetres. (10 dcm.)
= 100 centimetres. (100 cm.)
= 1,000 millimetres. (1,000 mm.)
1,000 Metres = 1 *Kilometre*.

CAPACITY.

1 Litre = 1 cubic decimetre. (1 c.dcm.)
= 1,000 cubic centimetres. (1,000 c.c.)

WEIGHT.

1 Gram or the weight of 1 c.c. of pure water at 4° C.
= 10 decigrams. (10 dg.)
= 100 centigrams. (100 cg.)
= 1,000 milligrams. (1,000 mg.)
1,000 Grams = 1 *Kilogram*.

ENGLISH EQUIVALENTS.

	NEARLY	ACCURATELY
1 Metre	= 3 feet 3⅜ inches or 39⅜ inches.	= 39·37079 inches.
1 Kilometre	= 1,100 yards	= 1093·6311 yards.
1 Litre	= 1¾ Impl. pint or 35 ounces.	= 1·76017341 pint.
1 Gram	= 15½ grains	= 15·43234 grains.
1 Kilogram	= 2¼ pounds	= 2·2046213 pounds.

MEMORANDA.

Weight of 1 litre of hydrogen (at 0° C. and under 760 mm. bar. pressure) = ·0896 gram, or 1 crith.
Therefore 1 gram of hydrogen measures 11·16 litres.
Specific gravity of Hydrogen (Air = 1) = ·0693.
,, ,, ,, Air (H = 1) = 14·42.

$$5° C. = 9° F.$$
$$\therefore 1° C. = \tfrac{9}{5}° F., \text{ and } 1° F. = \tfrac{5}{9}° C.$$

To convert C. into F. temperatures $\dfrac{C \times 9}{5} + 32 = F.$

To convert F. into C. temperatures $\dfrac{(F - 32)5}{9} = C.$

SECTION I.

CHAPTER I.

THE CONSTITUTION OF MATTER.

IN order to facilitate the explanation of chemical phenomena, modern philosophers have found it convenient to revive, in a somewhat modified form, the ancient hypothesis that all bodies possessing extension and weight are made up of stuff, substance, or *matter*, which is not uniform and continuous throughout, but consists of separate very small portions. Each of these small masses, which are called *molecules*, is supposed to be to a certain extent independent of the rest and isolated from them. The hypothesis further requires us to suppose that the molecules constituting any given species of matter are all alike, in size, weight, and properties, and differ in these respects from other molecules. Thus the molecules contained in one drop of water are conceived to be precisely like the molecules in any other drop of the same liquid. Similarly the molecules in a given globule of the liquid metal mercury, or quicksilver, must be assumed to be like all other molecules of the same metal; but water molecules and mercury molecules differ altogether from each other in weight and chemical properties. It must be distinctly understood that molecules cannot be seen, and all the arguments upon which the assumption of their existence is founded are derived from the examination of masses of appreciable magnitude. We know nothing of isolated individual molecules; the examination of these would be for obvious reasons impossible, and even if such a division were actually possible, the condition or properties of a single molecule would be in no way comparable with those of a mass of matter in which many molecules are imagined to be naturally aggregated by cohesion or otherwise. But we may assume that when one kind of matter affects chemically another kind, the smallest quantity of each which is capable of entering into the reaction consists of a determinate number of molecules. For the purposes contemplated in this book, then, a molecule may be defined as the unit of chemical action; that is, the smallest quantity which is able to take part in or result from a chemical change.

When two different kinds of matter are mixed together, chemical action often takes place between them. This action is supposed to be due to the union of molecules of the one kind with molecules of the other kind, or to an interchange of their components. The constituent parts of molecules [1] are called *atoms*.[2] Nothing is known concerning the true nature of atoms, though they have at different times been supposed to be hard particles of various geometrical forms, or vortices, like rings in air often produced from the top of a chimney, or from the funnel of a locomotive, and rendered visible by the smoke or steam. To the chemist the word simply means a small mass which cannot be divided, or, at least, has never been known to be divided, into smaller masses in the course of chemical action. An atom, then, is the smallest portion of matter which can be transferred from one molecule to another.

When the atoms within a molecule are all alike, the substance made up of such molecules is called an element. Most commonly, however, there are reasons for considering that the atoms composing a given molecule are dissimilar, and then the body is a compound. The number of atoms

---

[1] Molecule, diminutive, from Lat. *moles*, a mass.
[2] Atom, from α, not, and τέμνω, I cut.

in a molecule is very variable. A few elements have molecules which are assumed to consist of one atom only, others of two, three, or four; whilst in compound bodies the number may amount to hundreds.

Atoms are held together within the molecule, and probably molecules also are in certain cases united together by the operation of what is called, for want of a better term, *chemical affinity*, or *chemical attraction*. Of the nature of chemical attraction nothing is known. It cannot operate at a distance like gravitation, but requires that bodies should be in contact before they can combine together or act upon each other. It is apparently opposed by heat, for a sufficient elevation of temperature serves to break up the molecules of all compounds, and even those of many elementary substances, and in this disruption of molecular or atomic combinations by heat there is a very close analogy with the process of evaporation of a liquid. (*See* Dissociation.) It is closely related to electricity, and an electro-chemical theory, proposed by Davy in 1806, though never fully accepted, contains an element of probability which has led to its revival in a modified form in more recent times. This theory attributes chemical combination to the existence of charges of electricity of opposite kinds upon the atoms which unite, but there are several difficulties in the way which have never been satisfactorily cleared up, and are still a subject of debate. This question will be referred to again in a later chapter.

The study of those forces which affect entire molecules, and masses of molecules without regard to their composition, belongs to the domain of Physics. The immediate object of Chemistry is to ascertain the composition of bodies by the application of methods of *analysis* or of *synthesis*, but the science is not confined to the practice of these experimental arts. It has long been known that the same elements united in the same proportions do not necessarily produce compounds having in every case the same properties—in other words, the order in which atoms are united together is not less important than their number, and *constitution* must be clearly distinguished from *composition*. (*See* Isomerism.)

The density, fusibility, volatility, colour, crystalline form, and other physical properties of both elementary and compound bodies are dependent upon the nature, number, and order of the atoms composing their molecules, and it seems to be now fully established that the chemical characters of an elementary atom are directly related to its mass, as represented by the value of its atomic weight. Furthermore, the combination and the separation of atoms or of molecules is attended by a distribution of energy, and every chemical change is attended by absorption or evolution of heat, and frequently by other physical phenomena.

Chemistry is therefore a science of observation and of experiment, seeking by the study of the properties of bodies to establish a system of classification, and to connect that system with a knowledge of their constitution, both in regard to matter and the quantity of energy associated with it. When the science has made some further advances it may become possible more perfectly than at present to infer the constitution of a body from a knowledge of its properties, and on the other hand to predict with greater certainty the properties of the compound that would result from the union of given materials in any predetermined order.

The materials which compose the earth's crust, with its ocean and atmosphere and their inhabitants, may be roughly classified according to their mechanical condition into *solids* and *fluids*. Fluids are either liquids or gases.

A solid retains its form unless acted upon by pressure, by division with cutting instruments, by heat, or by solvents. Many solid bodies under suitable conditions assume definite geometric figures, which are generally bounded by plane faces, and thus give rise to crystals. This rigidity, by which the external form and relative position of parts is maintained, is due to what is called 'cohesion.' We should not, however, be justified in assuming that the constituent particles of solids are in a condition of absolute repose. Alteration of temperature and volume produced by the application of heat are attributable, according to the molecular theory, to the motion of the molecules, this motion being supposed to consist mainly in their rotation, or oscillation, or revolution, within a determinate space.

A liquid is recognised by its mobility, and by always assuming when at rest a horizontal level surface, except just where it comes into contact with the vessel containing it. The several parts of a mass of liquid are not held together with the same amount of force that binds together the parts of a solid, for it is generally much less difficult to detach a drop from a mass of liquid than to break off a portion of a solid. The curved surface of a liquid at contact with a solid, the ascent of liquids in capillary tubes, and the spheroidal form of the rain-drop as well as of water sprinkled upon a greasy surface, are all effects of 'surface tension'—that is, these phenomena are due to the fact that the superficial layers of a liquid behave like an elastic membrane stretched over the liquid and exerting a pressure upon it. The volume of a liquid is scarcely affected appreciably by pressure, even when very great, and pressure communicated to any one part of a mass of liquid is transmitted almost instantly and without loss to every other part.

From the phenomena of liquid diffusion which will be alluded to further on, it seems probable that the molecules of liquids are constantly moving about. These motions, however, are sluggish in comparison with the corresponding intestine movement which is observed in gases, and this is perhaps to be explained by the assumption that the molecules of liquids are comparatively close together, so that their free motion is impeded by frequent collision.

Gases differ from solids and liquids in the circumstance

that they seem to be entirely discharged from the influence of cohesion. A mass of gas exhibits no surface, like that of a liquid, and no gas can be confined in a vessel which does not enclose it on every side.

The volume of a gas increases as the pressure upon it decreases (Law of Boyle), until when the pressure is nothing the bulk of the gas becomes greater than the capacity of any conceivable vessel. If, therefore, a small quantity of a gas be introduced into any part of a vacuous space, it immediately spreads itself out and pervades every portion of that space equally.

Solids, liquids, and gases alike expand upon the application of heat, but whereas each solid and liquid increases by a fraction of its volume which is peculiar to itself, all gases expand to practically the same extent by equal increment of temperature. In other words, whilst the coefficients of expansion of solids and liquids are all different, those of true gases are the same in every case.

On the hypothesis that a gas, like a solid or a liquid, is a congeries of small masses or molecules, there are reasons for supposing that the molecules of gases are in constant and very rapid motion from place to place, and that they move in straight lines. A given molecule, however, cannot be supposed to pursue an uninterrupted course for any appreciable distance, but probably takes a new direction on the approach of another molecule.

These motions of translation, which, in accordance with received views, have been here attributed to the molecules of gases and liquids, are directly related to the amount of heat which has been expended in producing the liquid from a solid, or the gas from a liquid. In other words, when heat is consumed in melting a solid, or in gasifying a liquid, it is necessary to admit that the vibratory motion to which the temperature of the heating agent is supposed to be due, being communicated to the molecules of the object, may cause them not only to vibrate, but to move from place to place; and this altered motion corresponds with change of state.

For the further discussion of these speculations the reader must consult works on physics.[1] He will do well, however, constantly to set before his mind the fact that we possess at present no direct or positive proof even that molecules exist; still less have we any evidence regarding the conditions under which they may subsist in mass. The molecular hypothesis, however, is no longer in the position which it formerly held as a relic of the vague speculative philosophy of the ancients. It has been raised to the rank of a theory which bids fair to rival in completeness and importance the Newtonian theory of gravitation itself. In neither case does the theory admit of direct experimental proof; but both are accepted because they accord fully with the results of observation. The theory of molecules once admitted, all the recognised laws of chemical combination by weight and volume follow as necessary consequences. At the same time the phenomena connected with the physical properties of gases and liquids, such as the transmission of pressure and the remarkable laws of diffusion, find a rational and intelligible explanation such as no other hypothesis yet put forward has been competent to furnish. It is, in fact, not too much to assert that the rapid progress of modern chemistry and the intimate connection which it has been shown to have with other branches of physical science, as well as the illustrations it affords of the great doctrine of energy, are largely attributable to the general acceptance of this hypothesis. The science as it now stands may be regarded as a practical development of the molecular theory.

[1] The mathematical investigations of Krönig and Clausius, and of Rankine and Clerk Maxwell, have led to the establishment of the kinetic theory of gases, and have given a powerful impetus to the general recognition of the molecular theory by physicists. The student is recommended to make himself acquainted with the discussion of the subject from this point of view, which is given in Clerk Maxwell's *Theory of Heat*.

When a solid is transformed into a liquid, or a liquid into a gas, an apparently abrupt change of physical properties occurs, and heat is abundantly absorbed without producing elevation of temperature. But the transition from one state to another is by no means so sudden as appears from the consideration of cases like that of water. Many solids, such as iron, pass through an intermediate state, in which they are more or less plastic or viscid before they finally assume the liquid condition; and even the most perfect liquids with which we are acquainted are far from being absolutely mobile. Ether and alcohol, for example, flow more easily than water; but even these liquids exhibit a certain degree of viscosity.

Experiments commenced in 1822 by Caignard de la Tour, and since continued and extended by Dr. Andrews, have shown that matter is capable of existing in a somewhat analogous condition intermediate between the liquid and gaseous states:—' By partially liquefying carbonic acid gas by pressure and then raising the temperature to 88° F., the surface of demarcation between the liquid and gas becomes fainter, loses its curvature, and at last disappears. The space is then occupied by a homogeneous fluid, which exhibits, when the pressure is suddenly diminished or the temperature slightly lowered, a peculiar appearance of moving or flickering striæ throughout its entire mass. At temperatures above 88° F., no apparent liquefaction or separation into two distinct forms of matter could be effected, even when a pressure of three or four hundred atmospheres was applied.' (Andrews.) Nitrous oxide and sulphurous oxide, and other gases, give similar results. The striæ referred to are most probably the result of changes in density, caused by slight changes of temperature or pressure, as in ordinary liquids or gases when heated.

It thus appears that the various physical states of matter merge one into another by imperceptible gradations; and if we adopt the molecular theory we can see some explanation of this. The change of a solid into a liquid, and of a liquid into a gas, is the result of alteration, generally increase, in the distances between the molecules and in the rapidity and character of their movements. The generally received opinion is something like the following. In a solid every molecule is attached to the surrounding molecules by cohesion, and it is free to move only about a certain mean position, which it never leaves so long as the body remains solid. In a liquid each molecule clings to its neighbours, but less firmly than in the solid, and it is free to move and does move about in the mass, though proceeding in any one direction a very small distance before its path is altered by encounter with other molecules. In a gas the molecules are independent of each other, and move with various degrees of velocity in straight lines. When two molecules approach each other the path of each molecule is altered. In the intermediate states of matter there is probably a mixture of these conditions. Thus, in a liquid at its 'critical point,' like carbonic acid gas at 88° F., some molecules are probably moving singly and independently, like the molecules of a gas, in the midst of masses of others which are closer together and moving more slowly, after the manner of molecules in a liquid, the relative proportion of molecules in these two conditions serving to determine whether the mass exhibits the character of a gas or a liquid.

Experiments by Mr. Crookes on the phenomena exhibited by highly attenuated gases have led to the interesting discovery that when the tension of a gas is much reduced it possesses properties wholly different from those of the same gas under ordinary pressures. Under these circumstances, the number of molecules present in a given space being greatly diminished, the distance through which any one of them can move without collision is proportionately increased. When the pressure is reduced to about one-millionth of an atmosphere, and the discharge from an induction-coil is passed through the tube containing the gaseous residue, no luminosity is observed in the contents of the tube, as occurs when the electric discharge is passed through an ordinary vacuum, but the molecules driven from the negative pole with great velocity traverse the whole space before them in straight lines, and striking upon the surface of the glass or upon other solid bodies placed in their path, cause them to emit a phosphorescent light. The stream of molecules flowing thus from the negative pole is also capable of moving light bodies suspended in its course, and when brought to a focus and allowed to impinge upon the surface of a solid it heats it strongly. These phenomena are exhibited equally well by hydrogen, carbonic acid, or atmospheric air, and appear to have no connection with the chemical composition of the gas. It appears therefore that ordinary matter is capable of subsisting in a fourth state, differing as much from the solid, liquid, and gaseous states as these do from one another.

CHAPTER II.

FUSION.

WHEN, by the application of heat, a dry solid, such as sulphur or lead, is made to assume the liquid state, it is said to melt or undergo fusion. But when sugar is placed in water, it disappears, and is said to dissolve, and the liquid which results from such combination is called a solution. This distinction in terms is necessary, if only for practical convenience.

Every substance melts at a definite temperature, which is always the same for the same substance if pressure is constant. The slight changes of atmospheric pressure have no appreciable effect, but the influence of greater pressures may be recognised by the thermometer. Substances which, like ice, contract on liquefying, have their melting points lowered by pressure, whilst wax and sulphur, which expand on liquefying, have their melting points raised by pressure.

When fusion commences the application of heat causes no rise of temperature till the whole is melted.

It may be added that many substances, after being melted, require to be cooled below the melting point before they will solidify again, unless a particle of the solid is dropped into the liquid, when solidification at once commences and the temperature rises to the melting point. Even water, if kept quite still, may be cooled several degrees below zero without freezing. On shaking or dropping in a particle of ice it begins to crystallise and the temperature rises to $0°$, and remains there till the whole is frozen.

Amongst the elements the temperatures at which fusion occurs are very diverse, and there are but few cases in which any relation can be traced between this property and the chemical or other characteristics of the body. But among the compounds of carbon, which are very numerous, the determination of the melting point often serves as a convenient test whereby to distinguish two similar bodies from one another or to complete the identification of some substance under examination. As a general rule, it may, perhaps, be said that in a series of similar bodies those of smallest molecular weight melt at the lowest temperatures, as in the following examples :—

	Melting Point	Molecular Weight
Sulphur	$115°$	$n32$
Selenion	$217°$	$n79·5$
Tellurium	about $500°$	$n126$
Formic acid	$0°$	46
Acetic acid	$17°$	60
Palmitic acid	$62°$	256
Stearic acid	$69°·2$	284
Cerotic acid	$78°$	410
Melissic acid	$88°$	452

There are, however, numerous exceptions to this rule. Thus in the two following cases the melting points are in the inverse order of the molecular weights:—

	Melting Point	Molecular Weight
Cadmium	315°	112
Zinc	423°	65
Magnesium	low red heat	24
Potassium iodide	634°	166
Potassium bromide	699°	119
Potassium chloride	734°	76·5

In other series there is no direct relation between the molecular weight and the melting point of the substance. One example will suffice:—

	Melting Point	Molecular Weight
Barium nitrate	593°	261
Strontium nitrate	645°	211·5
Calcium nitrate	561°	164

In very many cases a mixture of two or more substances melts at a lower temperature than either of the ingredients.

Mixtures of the fatty and other acids melt at lower temperatures than the pure acids; the carbonates of potassium and sodium melt more easily when mixed than when alone; an alloy of potassium and sodium is liquid at the ordinary temperature; and an alloy of cadmium, tin, lead, and bismuth melts in hot water.

Further, it has been found that when two fusible substances are mixed together in certain proportions the melting point reaches a minimum—that is, the mixture melts at a temperature lower than the melting point observed when any other proportions are used. The following diagram represents the melting points of lead and tin and various mixtures of the same.

It will be noticed that whilst the melting point of tin is 228°, the temperature of liquefaction is lowered to 187° by mixing four atoms of tin with one atom of lead, although the melting point of the latter is considerably above that of tin—namely, at 326°. And upon trial it is found that if lead and tin are mixed together in the proportions of very nearly 2 parts of tin to 1 part of lead, or 3·3 atoms of tin to 1 atom of lead, the melting point is reduced to 180°, and no alloy containing the same metals in any other proportion melts at a

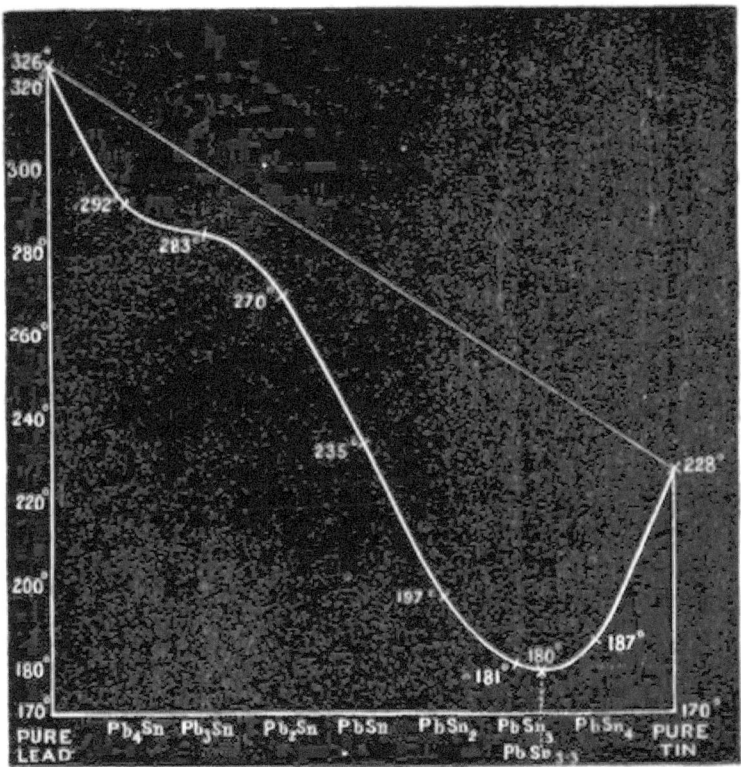

lower temperature. This sort of mixture of minimum melting point has been called a eutectic mixture, and the phenomenon has been called 'eutexia.'

There appears to be little doubt that the lowering of the freezing point of water by the addition of soluble salts is a phenomenon of the same order. It is well known, for example, that sea water does not freeze at the same temperature as fresh water, and a strong brine remains liquid at

20° below freezing point. Guthrie found that salts generally are capable of forming eutectic mixtures with water. Thus, if a solution of ammonium chloride, saturated at the ordinary temperature of the air, is cooled down, crystals of the anhydrous salt separate out. If these are separated from the solution as they are formed, the liquid on reaching a temperature of $-15°$ solidifies into a crystalline mass, which consists of 1 molecule of ammonium chloride united with 12 molecules of water. From this compound neither salt nor ice can be separated, and it melts and solidifies like a homogeneous mass. A compound of this kind was called by Guthrie a 'cryohydrate.'

The following are some further examples of these compounds:—

Cryohydrate	Temperature of Solidification
$ZnSO_4 + 20H_2O$	$-7°$
$MgSO_4 + 24H_2O$	$-6°$
$KNO_3 + 44H_2O$	$-2·7°$
$CuSO_4 + 44H_2O$	$-2°$
$KClO_3 + 222H_2O$	$-0·5°$
$K_2Cr_2O_7 + 292H_2O$	$-1°$

CHAPTER III.

SOLUTION.

Solution of Solids.—In discussing the questions relating to the phenomena of solution the following facts, among others, must be taken into account.

1. Considering any solid chosen as an example, we find that it is not soluble in all liquids. For example, common salt is readily and abundantly soluble in water, to a smaller extent in alcohol, and not at all in oils. On the other hand no liquid is capable of taking up every kind of solid. Water

dissolves sugar, common salt, nitre, and saline substances generally, but it has no appreciable solvent action upon many common metals and minerals, such as iron or copper, limestone or quartz, neither does it dissolve fat, oil, nor hydrocarbons generally.

Again, alcohol dissolves many carbonaceous substances, such as resins, phenols, acids, which are almost unaffected by water. Ether and benzene dissolve freely fats and solid hydrocarbons. Carbon bisulphide, which dissolves no salts, takes up many carbonaceous substances, and is the best solvent for common sulphur. Phosphorus trichloride and tribromide dissolve phosphorus. Fluid mercury dissolves many metals, but no other kind of substance.

There seems to be little doubt, in fact, that solubility is often dependent in some degree upon the existence of a similarity in composition between a solvent and the solid it dissolves. *Similia similibus solvuntur.* This is further illustrated as follows:—

Salts which contain water of crystallisation are, with comparatively few exceptions, easily soluble in water. Such compounds may be regarded as closely resembling water itself. A molecule such as $MgSO_4 \, 7H_2O$ may, in fact, be considered as a congeries of eight water molecules, in which one molecule of water is replaced by the elements of the salt. Again, the lower terms of the various series of alcohols and carbon acids show considerable similarity to water in their general behaviour, the higher terms much less. The lower terms of a series like the ethylic series of alcohols may be regarded as consisting of the elements of water, having one atom of hydrogen replaced by a hydro-carbon group, thus—

$$\left.\begin{array}{c}H\\H\end{array}\right\}O \quad \left.\begin{array}{c}CH_3\\H\end{array}\right\}O \quad \left.\begin{array}{c}C_2H_5\\H\end{array}\right\}O \quad \left.\begin{array}{c}C_3H_7\\H\end{array}\right\}O$$

These are miscible with water in all proportions. But when the hydrocarbon radicle becomes larger and more complicated, the resemblance to water and the miscibility with water is less. Thus—

$$\left.\begin{array}{c}C_4H_9\\H\end{array}\right\}O \qquad \left.\begin{array}{c}C_5H_{11}\\H\end{array}\right\}O \qquad \left.\begin{array}{c}C_6H_{13}\\H\end{array}\right\}O$$

are compounds of the same series as the foregoing, but they are not capable of mixing with water save in limited quantity. On reaching the highest known terms of the same series, we find waxy solids which are insoluble in water, but dissolve easily in ether or in hydrocarbon liquids, and themselves bear a strong resemblance to solid paraffin. The alcohol, $C_{16}H_{33}HO$, may be regarded as formed of the elements of the hydrocarbon, $C_{16}H_{34}$, in which H is replaced by HO, the latter, however, forming a very small proportion of the entire molecule. In the following series of compounds it will be noticed that the solubility in water increases as the amount of water residue, HO, present in the molecule increases.

Name and Formula.		Solubility in Water.
Benzene	C_6H_6	insoluble.
Phenol	$C_6H_5(HO)$	slightly soluble.
Catechol Quinol Resorcinol	$C_6H_4(HO)_2$	easily soluble.
Pyrogallol	$C_6H_3(HO)_3$	still more soluble.

2. The fusibility of a substance has much to do with its solubility. Thus, in comparing together three such very similar salts as chloride, bromide, and iodide of potassium, it is found that at all observed temperatures the most fusible, namely, the iodide, m.p. 634°, is more soluble in water than the bromide, m.p. 699°, and this again is more soluble than the chloride, m.p. 734°. However, it by no means follows that fusibility alone confers the property of solubility, for while sodium chloride dissolves abundantly in water, silver chloride, though much more easily fusible, is insoluble. The effect of fusibility is best observed when pairs of sub-

stances which are both soluble to some extent in a given menstruum are compared together at various temperatures.

3. This leads to another relation very generally observed, namely, that the solubility of solids usually increases with rise of temperature; in other words, a hot liquid dissolves a given solid more freely than the same when cold. Thus 100 parts of water dissolve—

Of	KNO_3	KCl	$KClO_3$	NaCl
At 0°	13·3	29·2	3·3	35·7
At 100°	247·0	56·5	56·5	39·8

These relations are best exhibited graphically in the following manner.

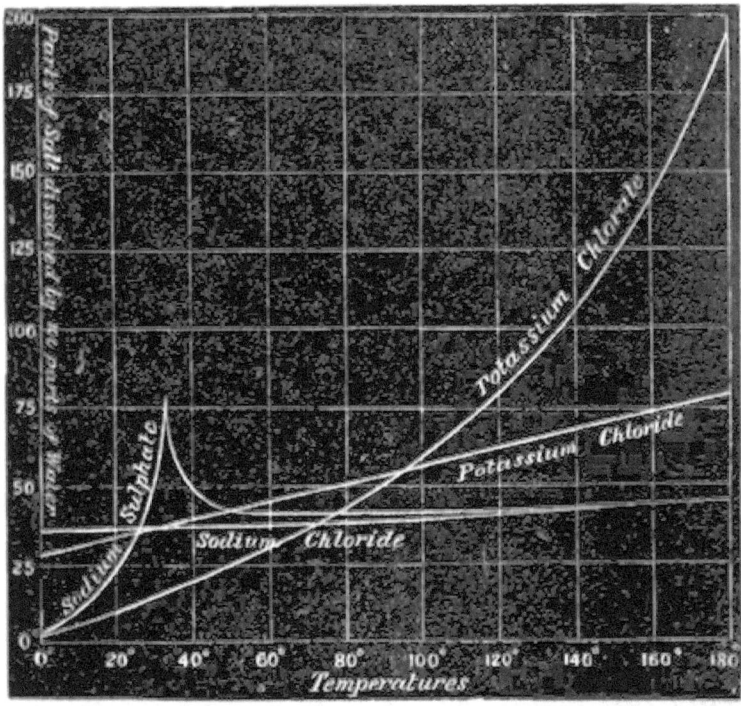

Here the divisions on the horizontal line represent temperatures, whilst those on the vertical express parts of salt dissolved at the several temperatures in 100 parts of water.

It will be noticed that the solubility of the easily fusible chlorate (m.p. 359°) increases rapidly with rise of temperature, whilst on the other hand the potassium chloride (m.p. 734°) and sodium chloride (m.p. 772°), which melt at higher temperatures, exhibit only a slight increase of solubility under the same circumstances.

To the rule that solubility increases with rise of temperature there are many apparent exceptions. Thus lime, calcium sulphate, and sodium sulphate are partly precipitated when a solution of each, saturated in the cold, is heated to boiling. The case of sodium sulphate is given in the diagram. It deserves to be noted, however, that the crystals which fall when a saturated solution of sodium sulphate is heated are anhydrous, whereas the crystals deposited at temperatures below 33° contain $10H_2O$.

4. When a solid dissolves in a liquid there is always a rise or a fall of temperature in the mass. Moreover, when the solution thus formed is mixed with more water there is again evolution or absorption of heat. The fall of temperature observed when such solids as common salt or nitre are dissolved in water is probably owing to the change from the solid to the liquid state. Many freezing mixtures, as of salt and snow, or sodium sulphate and strong hydrochloric acid, act by liquefying. On the other hand, when dry calcium chloride or sodium sulphate, or sodium carbonate dissolves in water, heat is produced, some of which, at least, is due to the chemical combination of the salt with a portion of the water.

In the following diagram is represented the thermal effect of adding successive doses of water in molecular proportions to sulphuric anhydride. The calorie is an amount of heat sufficient to raise the temperature of one part of water one degree. The divisions of the horizontal lines represent the number of water molecules added, and those on the vertical show the amount of heat evolved at any one stage, or in the entire process.

It will be seen that the thermal change becomes less and less as the liquid is diluted, but even when several hundred molecules of water have been added, the addition of one more produces an effect which can be recognised by the thermometer.

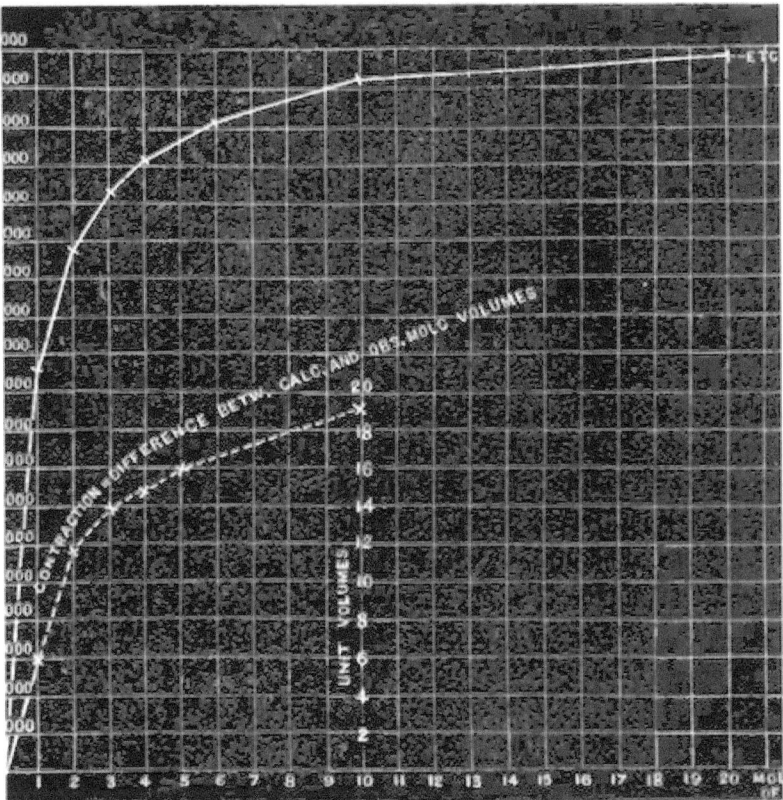

5. In the act of solution of solids, and especially of anhydrous salts in water, the volume of the solution is always less than the sum of the volumes of the solid and its solvent, with the exception of some ammonium salts, in which case it is greater. Similarly, the addition of water to a solution is followed by contraction.

The diagram already given exhibits the amount of contraction observed in the case of sulphuric acid and water, and the following table shows the observed and calculated molecular volumes by which the amount of contraction is indicated.

n	Molecular Volumes of $SO_3 + nH_2O$		Contraction = Difference between calculated and observed Mol. Vol.
	Density	Mol. Vol. = $\frac{\text{Mol. Wt.}}{\text{Density}}$	
0	1·940 (Solid.)	41·23	—
1	1·842	53·20	6·03
2	1·774	65·39	11·84
3	1·652	81·11	14·12
4	1·547	98·25	14·98
5	1·475	115·25	15·98
10	1·286	202·18	19·05

The contraction consequent on the first addition is greater than the second, and for each succeeding molecule is a diminishing quantity, becoming rather less than 1 after addition of $3H_2O$.

A similar result ensues if we calculate the contraction following upon the dilution of liquid H_2SO_4 with successive quantities of water.

The true explanation of this contraction is unknown, as also is the cause of the change of temperature always observed. But in both cases the effect is produced, not *per saltum*, but continuously, though in gradually decreasing amount as the water is added.

6. When a solid is dissolved in a volatile liquid the evaporation of the liquid is impeded, the pressure of the escaping vapour is reduced, and the boiling-point of the liquid is raised.

The table given on next page shows the effect of adding common salt to water.

A saturated solution of common salt under the pressure of one atmosphere boils at 108·4°.

Pressures of vapour evolved from pure water, and reduction of pressure on adding n molecules of NaCl to 100 H$_2$O at temperatures t°						
$t^\circ =$	70°	75°	80°	85°	90°	95°
	mm.	mm.	mm.	mm.	mm.	mm.
Water	233·3	288·8	354·9	433·2	525·5	633·7
$n = 2$	8·5	10·8	13·7	16·4	20·4	24·6
$n = 4$	18·0	22·5	28·0	33·8	40·9	50·2
$n = 5$	22·6	28·5	35·2	43·0	52·5	63·3

7. Coloured substances usually produce solutions of the same or very similar colour. But metallic salts which are capable of forming crystallised compounds with water often exhibit a variety of colours, according to the amount of water in which they are dissolved. Thus, dry copper sulphate is white, but it forms a blue solution in water. Cupric bromide is a black substance resembling iodine; a strong solution of it in water is dark brown, a weak solution pale blue. Cobalt chloride is blue, but when dissolved in water the liquid becomes successively indigo blue, violet, and finally red as more water is added. The light transmitted by such liquids corresponds very closely with that which passes through crystals of the salt containing various proportions of water. Thus, the salts just referred to are known to form definite compounds having a composition expressed as follows:—

Cupric sulphate.
$CuSO_4$ . . . white.
$CuSO_4\ 5H_2O$ . . blue.

Cupric bromide.
$CuBr_2$ . . . . black.
$CuBr_2\ 2H_2O$ . . blue.

Cobalt chloride.
$CoCl_2$ . . . . blue.
$CoCl_2\ 2H_2O$ . . indigo blue.
$CoCl_2\ 4H_2O$ . . reddish violet.
$CoCl_2\ 6H_2O$ . . red.

In order to account for these and other phenomena connected with solutions, various theories have been proposed. These are divisible into two classes, namely, (1) those which assume that a solution is formed as the result of chemical combination between the solid and the solvent, and (2) those which explain the phenomena by the adhesion of the solid to the liquid, and the mechanical intermixture of the two kinds of molecules.

The view which seems to conform most nearly to the results of observation is the following : A solid immersed in a liquid receives upon its surface the moving molecules of the liquid, and as a result some of the molecules from its surface may become detached and carried off into the liquid. If the encounter of the liquid molecules with those of the solid are incompetent to remove any of them, the substance is insoluble. Once in the liquid the molecules of the dissolved substance may be supposed to move about after the manner of liquid molecules, either singly or in groups, or in company with some of the molecules of the solvent. According to this, the process of dissolution is analogous to the process of evaporation, and just as in the case of evaporation we conceive that molecules which are near to the liquid surface may get entangled there and return to the liquid state again, so we must believe that, so long as a portion of the solid remains undissolved in the liquid, there is an interchange taking place between the liquid and solid molecules near its surface.

If temperature is raised the process of dissolution is usually promoted, for the molecules of both liquid and solid are more strongly agitated, and the cohesion between the particles of the solid is diminished. But for every given temperature there is a limit to the amount of solid dissolved (in all ordinary cases), when the number of molecules of solid passing off into the liquid no longer exceeds the number of molecules which return from the liquid to the solid surface and are retained there. Change of temperature must be believed to induce a change in the constitution of the liquid itself. In water near the freezing-point the molecules probably move about very sluggishly, and very few, if any of them, possess any degree of independence. But if the liquid is heated increasing numbers acquire the sort of movement which is proper to gas molecules, and increasing numbers are ready to pass off from the surface of the mass if they have an opportunity. If the water holds a substance in solution, the molecules of this substance, like those of the solvent, tend to aggregate together at lower temperatures, and to separate from one another and from the molecules of solvent at higher temperatures.

The question arises, what is it which causes salt molecules to unite with water molecules in a saline solution? The partisans of one view say it is chemical attraction or affinity; the partisans of the other prefer to call it adhesion. Whatever name be given to it, however, it differs in no obvious quality from the 'attraction' which subsists between the salt and the water in solids which contain 'water of crystallisation.' (See also Chapter XXII.)

Naturally connected with the production of solutions is the process of crystallisation, as it is in the gradual passage from the liquid to the solid state that we find the most generally useful method for the formation of crystals.

The crystallisation of a crystallisable solid from its solutions occurs when the percentage of solid present in the liquid exceeds a certain limit dependent on the nature both of the solid and the liquid in which it is dissolved. This condition may be brought about either by allowing part of the solvent to evaporate, or by altering, in general by lowering, the temperature. Crystals are then formed, and these crystals, in the majority of cases, contain not only the elements of the dissolved substance, but a portion of the solvent united to it in definite molecular proportions. This is particularly noticeable in the case of aqueous solutions, and the water thus combined is spoken of as *water of crys-*

tallisation. Alcohol and benzene, and probably other liquids, unite with salts and other crystallisable bodies in the same manner. It is noteworthy that the proportion of water of crystallisation is principally dependent upon the temperature at which the process of crystallisation takes place. Thus, sulphate of sodium crystallises from water at temperatures above 34° in the anhydrous state. But at the ordinary temperature of the air, the solution deposits crystals which contain ten molecules of water with one molecule of the salt.

Certain solutions refuse to crystallise when cooled down below the temperature at which, under ordinary circumstances, they yield crystals. Such liquids are said to be *supersaturated*. In order to prepare a supersaturated solution, it is only necessary to warm some water gently in a test-tube or flask and add to it crystals of sodium sulphate, or of alum, or sodium carbonate, so long as they are dissolved. If the solution is then filtered clear, covered to protect it from dust, and allowed to cool, it will generally remain liquid. If now a small particle of the same salt be dropped into the liquid, crystallisation at once commences and proceeds so rapidly that in a few minutes the whole sets into a solid mass. In most cases crystallisation is not started by shaking the liquid, nor by the introduction of a crystal of a different salt, unless it be strictly isomorphous with the dissolved substance.

Solution of Gases.—All gases dissolve to a greater or less extent in water, but, unlike solids, their solubility diminishes as the temperature rises, so that in most cases [1] the dissolved gas may be completely expelled from a liquid by boiling, whilst the amount taken up may be greatly increased by cooling the liquid.

Increase of pressure also augments the solubility of gases in a direct ratio. It is therefore necessary in making any

---

[1] Exceptions occur in the cases of hydrochloric acid and some other gases.

statement as to the solubility of a gas to observe the conditions of temperature and pressure under which that solubility was estimated. The following examples will serve to show how greatly gases differ in the extent to which they dissolve in water.

At 0° C. and under a pressure of 760 mm. barom.
1 volume of water will dissolve

Hydrogen	·0193
Nitrogen	·02035
Oxygen	·04114
Nitrous oxide . . .	1·3052
Carbon dioxide . . .	1·7967
Hydrogen sulphide . . .	4·3706
Sulphur dioxide . . .	79·789
Ammonia	1148·8

The numbers given above represent volumes of the several gases measured at 0° and 760 mm., and constitute the *coefficients of absorption* of these gases at that particular temperature and pressure.

The general statement that the weight of a gas dissolved by a liquid is directly proportionate to the pressure is often known as the law of Henry and Dalton. It admits of another expression; for since, according to the law of Boyle, the volume of a gas diminishes as the pressure upon it increases, it is obvious that the volume of gas thus held in solution must always be the same, whatever the pressure.

These rules no longer hold good when the gas and the liquid exert a chemical action upon each other, and exceptions must also be recognised in the case of the more soluble gases, such as hydrochloric acid and ammonia.

The determination of the absorption of gases by liquids may be applied in certain cases to the elucidation of some important theoretical and practical questions. Atmospheric air furnishes an example which will be worth the consideration of the student.

When a mixture of gases is exposed to the action of a solvent, the quantity of each of the constituents dissolved by the liquid will depend first upon its coefficient of solubility, and secondly upon the proportion in which it exists in the mixture. This proportion determines the pressure which each gas present exerts upon the surface of the liquid, and consequently regulates the amount of it which is dissolved. The total pressure produced by the mixture is therefore the sum of those partial pressures due separately to the individual constituents. To make this more clear:

Suppose a very large vessel containing a very little water, and filled with oxygen, under a pressure of one atmosphere. It is plain that if four-fifths of the gas were removed the pressure would be reduced to one-fifth of an atmosphere, and the quantity of oxygen dissolved would be only one-fifth the quantity taken up under the previous conditions, provided, of course, that the temperature remain constant. An exactly similar vessel can be conceived filled with nitrogen under one atmosphere, and containing a little water. If one-fifth of the nitrogen were removed, the pressure of the remainder would be only four-fifths of an atmosphere, and the quantity dissolved would be reduced to four-fifths. Lastly, a similar vessel, filled with atmospheric air, contains a gas in which the conditions of the two previously supposed experiments are combined.

Air is composed very nearly of four volumes of nitrogen to one volume of oxygen, and by reason of the greater solubility of oxygen the proportion of the two gases one to the other is found to be disturbed when air is shaken up with water, the dissolved gas being richer in oxygen, the residual air richer in nitrogen, than the original.

No stronger evidence could be adduced in favour of the view generally held, that in atmospheric air the two main components are not united chemically, but are in a state of intimate mechanical mixture.

Note.—The following is an example of the kind of problem that might occur in connection with this subject :

Calculate the percentage composition of the gas which would be dissolved by water exposed in a room full of air containing 79 N, 20·6 O, and ·4 of CO_2 in 100 volumes (temp. 0° and bar. 760 mm.).

Coeff. of sol. for oxygen . . . ·04
„ „ nitrogen . . . ·02
„ „ carbon dioxide . . 1·79

The pressures are proportional to the volumes of the gases present. Therefore the relative quantities dissolved would be :

Nitrogen   79   ×   ·02   =   1·58
Oxygen   20·6 ×   ·04   =   ·824
Carbon dioxide ·4 ×   1·79 =   ·716

The total quantity . = 3·120

The percentage composition of the dissolved gas would therefore be :

Nitrogen . . . . . . 50·6
Oxygen . . . . . . 26·4
Carbonic dioxide . . . . 22·9

CHAPTER IV.

LIQUID DIFFUSION AND DIALYSIS.

An aqueous solution of sugar or of salt is heavier than water, and may be readily poured through a funnel with a long stem into a glass of water in such a way as to form a separate stratum at the bottom. If the solution is coloured it will soon be noticed that the colour gradually extends upwards through the liquid, until, after a few hours or a few days, according to circumstances, the whole liquid is tinged. But it is not necessary that the liquid should be coloured. The taste, specific gravity, refractive power, or the application of chemical tests will soon give indications that the solution from below is mixing with the liquid above. This process of spontaneous intermixture is called diffusion. It results from the proper motion of the molecules of the liquid, and cannot be referred to the disturbing influences of changes of temperature. The rapidity with which diffusion of this kind takes place, and the limit of its action, depend very much upon the nature of the liquids employed.

The power of interdiffusion is by no means universal among liquids, some liquids being, like mercury, oil, and water, quite incapable of mixing together under any circumstances; whilst others, such as water and solution of hydrochloric acid, mingle spontaneously in consequence of very rapid diffusion. We are indebted for nearly all the information we possess on this subject to the late Professor Graham. His experiments were conducted very nearly in the manner already described at the beginning of the chapter. The glass vessel in which diffusion was allowed to go on was graduated into equal divisions, from the bottom upwards; and after the introduction of the two liquids the whole was left in a room, the temperature of which was kept as uniform as possible.

After a time, the liquid occupying successive divisions of the vessel was removed by a small syphon or pipette and analysed, in order to ascertain the extent to which diffusion had taken place. In this way a number of conclusions were arrived at, amongst which the following are the most important:—

1. Bodies are divisible, as regards their diffusive power, into two classes. Those which diffuse most readily through a given liquid menstruum are, for the most part, crystallisable substances, and are termed by Graham *crystalloids*; whilst the least diffusible bodies are uncrystallisable, with, in most cases, high molecular weight, and are denominated *colloids*,[1] from their resemblance to glue, which may be taken as the type of this class. The following list supplies the *times* of equal diffusion by the substances there named; and

---

[1] κόλλα, glue.

it will be seen that albumen and caramel, both of which are uncrystallisable substances of somewhat indefinite composition, are far behind the rest :—

Hydrochloric acid	1
Chloride of sodium	2·33
Sugar	7
Sulphate of magnesium	7
Albumen	49
Caramel	98

Thus hydrochloric acid diffuses more than twice as rapidly as chloride of sodium, seven times as rapidly as sugar or magnesium sulphate, forty-nine times as rapidly as albumen, and nearly one hundred times as rapidly as caramel. Hydrochloric acid is one of the most diffusive substances known.

2. Equal rates of diffusion are exhibited in many cases by the members of isomorphous groups. Thus hydrochloric, hydrobromic, and hydriodic acids have nearly the same diffusion rate ; so also have the chlorides, bromides, and iodides of the alkali metals ; the nitrates of barium, strontium, and calcium, and the sulphates of magnesium and zinc.

3. The rate of diffusion increases with the temperature, and when the solution is not too concentrated is proportional to the strength of the solution.

By taking advantage of this difference in diffusibility, mixed salts may be separated from one another to a certain extent, and crystalloids may be isolated pretty perfectly from admixture with colloids. In the practical application of this process it has been found convenient to separate the liquids undergoing diffusion by some membrane or partition composed of colloid material, and this mode of diffusion, through a septum, is called *dialysis*. The process is a very simple one. The liquid holding in solution a mixture of crystalloids and colloids is placed in a sort of tray or sieve, formed of a sheet of parchment paper stretched over a hoop. This vessel, which is called the *dialyser*, is made to float in a dish of pure water, which after a time can be renewed if necessary. Under these circumstances, the crystalloids pass out by diffusion through the membrane (which must be perfectly free from holes), and by evaporating the liquid down may be obtained in a condition of tolerable purity.

The application of this method led to the discovery of the soluble colloidal forms of ferric hydrate, silicic acid, alumina, and other bodies which had been previously known only in the pectous or gelatinoid condition, and the study of which could not fail to throw considerable light on the obscure natural processes by which these bodies are deposited as minerals in the crystalline form, and in a great state of purity. If, for example, we take a solution of silicate of sodium, and add to it a slight excess of hydrochloric acid, we obtain a perfectly clear liquid, which contains the very substance referred to above as colloid silicic acid; but in this liquid it is mixed with the acid used and the common salt formed by the decomposition. The process of dialysis furnishes the means of separating these latter substances without causing the precipitation or other alteration of the silicic acid, which is left on the dialyser in the form of a colourless limpid solution. This solution is, however, very unstable, especially when concentrated, and the addition of even very minute quantities of various salts causes the whole of the silica to separate out in the form of a translucent jelly, which cannot be re-dissolved, except by the addition of a fresh quantity of alkali.

Liquid stannic, titanic, tungstic, and molybdic acids have been prepared by a similar process.

The ultimate pectisation of liquid silicic acid and other colloids is preceded by a gradual thickening of the liquid, and just before gelatinising silicic acid flows like an oil. These effects are doubtless the result of the tendency of the particles of colloids to cohere, aggregate, and contract.

This tendency manifests itself occasionally in the exercise of very considerable force. Thus the contraction of gelatine drying in a glass dish over sulphuric acid, together with the adhesion of the gelatine to the glass, is said to be sufficiently powerful to tear up the surface of the glass. Glass is itself a colloid, and the permanent adhesion between the surfaces of polished plates of glass left in contact with each other, which is said to occur sometimes, is referred by Graham to this class of phenomena.

CHAPTER V.

EVAPORATION AND EBULLITION.

A LIQUID boils when, by raising its temperature, the elasticity of the vapour formed at any point in the liquid is capable of overcoming the pressure at that point. This pressure is made up of the pressure of the atmosphere and that of the superincumbent stratum of liquid. In a vessel of inappreciable depth, and when the mercurial column in the barometer is 760 mm. high, alcohol boils at $78°·4$ and water at $100°$.

The tension of alcohol vapour is, therefore, at any temperature below its boiling point, greater than that of water, and alcohol is said to be more *volatile* than water. Nearly all liquids are volatile, but the temperatures at which they evaporate freely are very diverse. There is, for example, a wide range between the volatility of ether and of molten silver, or between that of liquid carbon dioxide and of metallic mercury.

The observation of the boiling point of liquids is an operation of daily occurrence in the laboratory, and although very few general laws connecting the boiling point with chemical characters have been traced out, a few

D

general observations have been made, to which the attention of the student must be directed. The conversion of a liquid into a gas or vapour is attended by the absorption of heat, and this heat is consumed in giving to the molecules of the liquid a new and more rapid motion. From this consideration it would appear that bodies formed of light molecules would be more easily vaporised than others constituted of complex and consequently heavy molecules. Such is indeed the case, and, in the broadest sense, such a statement would be nearly true. But when we examine individual cases, we meet with so many exceptions and anomalies, that it is obvious such a law must be applied with extreme caution. It is, however, permissible to say that bodies which are strictly comparable in regard to chemical and other physical qualities do in nearly all cases exhibit the relation referred to. Among the elements the most noteworthy instances are the following :—

	Molecular Weight.	Boiling Point.
Chlorine	71	$-50°$
Bromine	160	$63°$
Iodine	254	$175°$
Oxygen	32	gaseous
Sulphur	64?	$440°$
Selenion	159?	below a red heat
Tellurium	252?	white heat
Nitrogen	28	gaseous
Phosphorus	124	$290°$
Arsenic	300	volatilises at red heat without fusion
Antimony	240?	white heat
Bismuth	416?	white heat

Groups of similarly constituted compounds show the same relations.

Boiling Points.

	Molecular Weight.	Boiling Point.
Sulphur Dioxide	64	$-10°$
Sulphur Trioxide	80	$46°$
Chloromethane	50·5	gaseous at ordinary temperatures
Dichloromethane	85	$30·5°$
Trichloromethane	119·5	$61°$

But the most important cases are observable among the carbon compounds, which form *homologous* series, to which reference will be made hereafter. One example will suffice in this place. The following compounds all contain carbon and hydrogen in proportions which may be represented by the general formula C_nH_{2n+2} :—

	Molecular Weight.	Boiling Point.
Methane (marsh gas)	16	gaseous
Ethane	30	gaseous
Propane	44	gaseous
Tetrane	58	$1°$
Pentane	72	$38°$
Hexane	86	$70°$
Heptane	100	$99°$
Octane	114	$124°$
	&c.	&c.

In this series, for every increase of 14 on the molecular weight, the boiling point rises by about $30°$ to $35°$.

In the determination of boiling points it is usual to observe, by an accurate thermometer, the temperature of the vapour evolved by the boiling liquid, and not that of the liquid itself. This precaution is necessary when glass vessels are, as usual, employed, in consequence of a peculiar adhesive attraction which glass exercises, and which causes the boiling point to be slightly raised above the true temperature of ebullition. This adhesion, assisted, doubtless, by cohesion in the liquid itself, sometimes gives rise to the phenomenon of irregular ebullition or 'bumping.' The entire thread of mercury in the thermometer should be immersed in the vapour, so as to be heated by it, and the height of the barometer should be noted at the time of the experiment.

This last is a precaution which is very generally neglected, and its neglect is probably the cause of some of the discrepancies noticed between calculated and observed boiling points and between the results of different experiments.

The determination of the boiling point is often useful in deciding as to whether a given liquid is a mixture or a homogeneous body. When heat is applied to a mixture of volatile liquids the mixture begins to boil at a temperature very near to the boiling point of its most volatile constituent, and if the temperature is not allowed to rise above this point, ebullition in most cases soon comes to an end. But if the application of heat is continued whilst the thermometer is kept in the vapour, the temperature may be observed to rise continuously till the whole of the liquid has boiled away. If this operation is conducted in a flask connected with a condensing apparatus and a receiver, and if the receiver is changed at intervals, so that the several portions which pass over between certain limits of temperature are received in separate vessels, a more or less complete separation of the constituents of the liquid may be effected. Such a process is called *fractional distillation.*

The compression of a vapour tends to produce the same change of state as lowering its temperature. In either case the approximation of the molecules is attended sooner or later by the liquefaction of a part of the vapour. With these facts in view, and considering the generally close resemblance between vapours and those bodies which are commonly called true gases, Faraday came to the conclusion that the latter are not essentially different from the former, but are in truth the vapours of volatile liquids far removed at ordinary temperatures from their boiling points. This conclusion he verified experimentally by enclosing in strong $\wedge$-shaped glass tubes materials capable of evolving the gases he wished to examine. On the application of a gentle heat to these materials gas was generated, and by its accumulation in the confined space, sufficient pressure was exerted to cause its partial liquefaction. In this way ammonia, chlorine, and other gases were reduced to the condition of limpid liquids, and by the combined use of pressure and low temperature, produced by powerful freezing mixtures, a great many other bodies, which till then had been known only in the gaseous form, were also liquefied. Larger apparatus, constructed on the same principle as Faraday's glass tubes, were subsequently employed by different experimenters, and at the present time several gases, such as carbonic anhydride and nitrous oxide, are liquefied on a large scale by compressing them by powerful force-pumps into iron cylindrical bottles fitted with stopcocks. By such methods all known gases were, with six exceptions, reduced to the liquid state.

It was not until the end of the year 1877 that these exceptions disappeared, and it was found possible by application of enormous pressure assisted by very low temperature to reduce to the liquid state such gases as oxygen, hydrogen, nitrogen, and atmospheric air. This interesting result was arrived at almost simultaneously, though independently, by M. Pictet of Geneva and M. Cailletet of Chatillon-sur-Seine. A temperature of $-130°$ combined with a pressure of about 475 atmospheres is required for the liquefaction of oxygen, whilst at about the same temperature a pressure of some 650 atmospheres is necessary in the case of hydrogen.

The evaporation of liquefied gases is attended by the absorption of much heat, and in some cases the reduction of temperature is such as to cause the solidification of part of the liquid. Carbonic anhydride can be obtained in the form of a white snow-like solid by allowing a fine stream of the liquefied gas to escape into the air. Part of it evaporates very rapidly, and so much heat is thus rendered latent that the remainder freezes. Hydrogen is also said to yield a solid, though in this case, owing to the difficulties attending the experiment, there is some doubt about the observation.

The following table shows the amount of pressure in atmospheres necessary at the temperature of 0° to liquefy some of the more important of the gases :—

PRESSURE IN ATMOSPHERES—TEMP. 0° C.

Sulphur dioxide	1·53
Cyanogen	2·37
Hydriodic acid	3·97
Ammonia	4·4
Chlorine	about 4
Hydrogen sulphide	10
Nitrous oxide	32
Carbon dioxide	38·5
Hydrochloric acid	about 42

Some of these may be liquefied by cold alone under the ordinary atmospheric pressure. Thus—

Sulphur dioxide condenses at	−10°
Cyanogen ,,	−22°
Ammonia ,,	−36°
Chlorine ,,	−50°
Carbon dioxide ,,	−87°

On the other hand, Andrews found (see also p. 10) that if the temperature be raised to a certain point, a gas which is otherwise liquefiable can no longer be liquefied by pressure. This point, which varies with the nature of the gas operated upon, is called the *critical point*. In the case of carbon dioxide it is 31° C. The critical points for hydrogen, oxygen, and the rest of the gases which were formerly supposed to be permanent are at temperatures so low as to be attainable only by special arrangements and with great difficulty. Hence the failure of the earlier attempts to reduce these gases to the liquid state, notwithstanding the application of pressures represented by several thousand atmospheres.

It has been already stated earlier in the chapter that if we compare together liquids which consist of the same elements, and which present the same general properties, those which have the simplest constitution are the most readily converted into vapour. The converse is also, to a certain extent, true. Comparing together similarly constituted gases and vapours, we find that those which are composed of simple molecules are often more difficult to liquefy by cold or compression than others of more complex constitution. These facts are sometimes serviceable in helping to decide questions as to the relative complexity of two nearly allied compounds. For example, there are two oxides of carbon, both gaseous at ordinary temperatures, but one of them more easily capable of liquefaction under pressure than the other. The more easily liquefiable oxide is in all probability made up of heavier molecules than the other, and this view is supported by a comparison of the densities of the two gases. Carbonic anhydride gas is, bulk for bulk, 1·57 time heavier than carbonic oxide.

Nitrous and nitric oxide, ethylene and marsh gas, furnish examples of the same kind.

Such a rule, however, is liable to many exceptions, and can only be used in conjunction with other considerations. Thus ether, $C_4H_{10}O$, is more volatile than alcohol, C_2H_6O. Moreover, it often happens that the same gas or vapour changes in constitution at different temperatures. For example, nitric peroxide above 180° is composed of molecules whose formula is NO_2. At lower temperatures these combine in pairs to form more complex molecules, N_2O_4, so that when the gas is about to liquefy it consists of molecules of greater mass than when it is hot. Hydrofluoric acid affords another example of the same kind. Its molecule consists of HF at moderately high temperatures, whilst at a little above 20° it becomes H_2F_2. It is probable that there are many cases of this kind. (See Chapter XVIII., 'Dissociation.')

CHAPTER VI.

DIFFUSION AND DIALYSIS OF GASES

A VERY remarkable property of gases and vapours is their power of mixing with one another, even in opposition to gravity.

If a bottle of any odorous gas is opened in any part of a room of constant temperature and free from draughts, the smell of the gas soon becomes perceptible in every part of the room, and, after the lapse of a short time, equally in every part. Notwithstanding that, as in the case of sulphuretted hydrogen, a heavy gas may be selected for the experiment, it would be easy to prove by analysis of the air that every part of it is equally impregnated with the foreign matter. Other experimental illustrations of the same law may easily be devised. A bottle of hydrogen held mouth downwards in the air for a short time soon becomes explosible. A jar of air inverted over another filled with carbonic acid gas soon acquires the power of giving, like carbonic acid, a precipitate with lime water. And this process of intermixture proceeds almost equally well if the gases are separated from each other by a partition formed of some porous material.

A thin plate of unglazed earthenware, a slice of artificially compressed graphite, or a cake of dry plaster of Paris, may be employed for the purpose. A very effective form of apparatus consists of a clay battery cell closed by a cork, through which a yard or so of glass tubing open at both ends is made to pass. By means of this simple apparatus it may be shown that different gases penetrate the porous clay with different degrees of rapidity, and that light gases effect a passage more quickly than heavier ones. The consequence of this difference of diffusion-rate is that a difference of pressure is established inside the cell, and if the open end of the tube is dipped into water a certain quantity of gas is expelled in bubbles from below, or the liquid is forced by the atmospheric pressure up the tube. If, for example, the clay vessel previously full of air is surrounded by hydrogen gas, intermixture of the air within and the hydrogen without takes place through the clay; but since the hydrogen diffuses more rapidly than the air, the quantity of gas within is rapidly increased, and some of it visibly finds its escape from the open end of the tube through the water.

Early observations of and experiments upon gaseous diffusion were made by Priestley in the last century, and by Döbereiner in 1825; but Graham gave the explanation of the phenomena, and by precise and long-continued experiments established the law:

The velocities of diffusion of different gases are inversely proportional to the square roots of their densities.

Graham's experiments were for the most part conducted with a very simple apparatus, consisting of a straight wide glass tube closed at its upper extremity by a disc of porous stucco or graphite. This tube was filled with hydrogen or other gas over the mercurial trough, the graphite plate being covered during this operation with a sheet of guttapercha. The mercury within the tube was kept at the same level as the mercury in the trough, in order that there might be no alteration of pressure whilst the diffusion was proceeding, and at the same time the temperature and barometric pressure were recorded. After the lapse of a certain interval, measured by a chronometer, the volume and composition of the residual gas could be determined.

The following table embodies some of the results

obtained in this way, the pressure and temperature being supposed to be the same in all cases:—

DIFFUSION OF GASES.

Name of Gas	Density	Square Root of Density	$\dfrac{1}{\sqrt{\text{Density}}}$	Velocity of Diffusion
Air	1	1	1	1
Hydrogen	·0693	·2632	3·7794	3·83
Marsh gas	·554	·774	1·3375	1·344
Carbonic oxide	·9678	·9837	1·0165	1·0149
Nitrogen	·9713	·9856	1·0147	1·0143
Oxygen	1·1056	1·0515	·9510	·9487
Carbonic dioxide	1·529	1·2365	·8087	·812

It will be noticed that the observed rate of diffusion agrees very nearly with the rate calculated from the density of the gas; but in no case is there absolute concordance. This is, probably, in part due to the errors inevitable in any experimental investigation, especially where gases are concerned, but is also in some degree attributable to the fact that the diaphragm employed possesses an appreciable thickness, so that in passing through the pores the gas encounters considerable resistance.

By taking advantage of the unequal diffusibility of gases of different density, a partial separation of mixed gases may in some cases be effected.

The gases constituting atmospheric air, for example, may be, to some extent, separated from each other by causing a slow current of air to flow through a clay tube passing through a glass tube which has been exhausted as completely as possible by the air-pump. The nitrogen, being lighter and consequently more diffusible than the oxygen, passes more abundantly into the vacuous space, leaving the residual air richer in oxygen than it was originally.

The physical explanation of the phenomena of diffusion depends directly upon the mechanical theory of gases, which has been (Chap. I.) already discussed. The molecules of a gas are supposed to move constantly in straight lines till they come nearly into contact with other moving molecules, or with the walls of the containing vessel. If these walls are perforated at intervals with apertures large enough to permit the passage of a molecule, we may conceive that although many molecules continue to rebound as though the surface were impervious, yet a great many others may find their way into and through these short passages, and so into the atmosphere beyond. Molecules from the external atmosphere may be assumed to pass inwards in precisely the same manner, and if the densities of the gases on the two sides of the partition are the same, the number of molecules passing inwards in a given time is exactly equal to the number passing outwards, and no change of volume or of pressure can result. But if the gases are of different densities, molecules of the lighter gas pass through more rapidly than those of the heavier, and a change is produced in the tension or elastic force of the gas enclosed in the porous vessel.

This pressure, exerted by a gas in opposition to that which it has to bear when enclosed in a vessel, is represented according to the kinetic theory as the result of the continuous showering down of its molecules upon the surfaces with which it is in contact. The molecules of different gases being of different masses, they must move with different degrees of velocity, the light molecules more rapidly than heavier ones, in order to produce the same amount of pressure. Hence, when the gas in contact with the porous surface is a light gas, its molecules must be supposed to fall upon a given area more frequently than when the gas employed is heavier, and consequently the opportunities for the escape of molecules through the pores are more frequent. Hence it is that light gases diffuse more rapidly than heavy gases.

Gases have not only the power of passing by diffusion

through porous substances, but under certain circumstances penetrate membranes, and even sheets of metal which are absolutely destitute of pores. This phenomenon differs entirely from diffusion, for it is not found that the lightest gases traverse such substances most rapidly; indeed, the contrary is more generally the case. Moreover, the metals which are so remarkable for their power of transmitting some gases are absolutely impermeable by others.

A few simple experiments will give the student an idea of the general character of the phenomena we are discussing. If a thin india-rubber balloon (such as are sold at the toy-shops), inflated with air, is immersed for a few minutes in a vessel full of carbonic dioxide gas, the balloon becomes largely distended, and if a band of tape is fastened round it before the experiment, it generally bursts after immersion in the gas for a short time. A similar balloon filled with hydrogen or carbonic dioxide gas quickly collapses when exposed to the air. Such a film of rubber appears to have no porosity, but rather to resemble a film of liquid in its relations to gases. The penetration of the rubber and similar colloids by a gas appears to be due to the absorption of the gas by one surface of the colloid and its transmission to the other surface by the agency of liquid and not gaseous diffusion. The liquefied gas then volatilises into the vacuum or atmosphere on the other side.

The passage of gases through metallic plates at a red heat is referred by Graham to a somewhat similar cause. Thus, at a red heat, both platinum and palladium, and even iron, are permeable by hydrogen gas; and this is evidently connected with the fact that the same metals are capable of absorbing and retaining considerable quantities of hydrogen when that element is presented to them under suitable conditions. Thus a sheet of palladium connected with the negative pole of a battery, and immersed in acidulated water, becomes charged with upwards of 200 times its volume of hydrogen gas. The same metal in a spongy state absorbs

686 times its volume of hydrogen when heated in the gas to 200°. This 'occlusion' of hydrogen is not attended by any alteration in the appearance of the metal, although its volume is increased, and consequently its density diminished.

Other gases are occluded in a similar manner by other metals; but in each case a certain selective power is manifested on the part of the metal. Thus, platinum and palladium take up hydrogen freely, but no other gas to an appreciable amount; iron takes up hydrogen and carbonic oxide, and melted silver absorbs oxygen.

It is not improbable that these remarkable effects may be special manifestations of the absorptive power which is possessed in greater or less degree by the surfaces of all solids. Even glass exposed to air condenses a quantity of gas which extends, in the form of an invisible and almost imponderable film, over the entire surface. The gas thus collected from air by glass consists of the same ingredients as are found in common air, namely, oxygen, nitrogen, carbon dioxide, and water; but it is not quite certain whether they are present in the same proportions. If the surface is greatly extended, as is the case with porous substances like plaster, earth, and especially charcoal, the amount of gas taken up is often very large, and each substance seems to possess a power of selection from the gases presented to it. Cocoa-nut charcoal, for example, at the common temperature and pressure of the air, absorbs about 170 times its bulk of ammonia, but only about 73 times its bulk of carbon dioxide, and a very much smaller quantity of oxygen, nitrogen, or hydrogen.

As a general rule those gases which are easily liquefiable are absorbed by such substances in largest quantity.

CHAPTER VII.

RELATION OF GASES TO TEMPERATURE AND PRESSURE.

THE bulk of a given portion of matter, solid, liquid, or gaseous, must depend both upon the temperature and the pressure to which it is submitted. The variation of volume produced in a solid or liquid by change of pressure or of temperature is very small when compared with the change produced by the same means in the volume of a gas. The law of the compressibility of gases was discovered by the Hon. Robert Boyle and published in 1662. It was examined independently and verified by Mariotte some years afterwards.

Law of Boyle.—'The volume of a given mass of any gas varies inversely as the pressure.'

Thus, if V is the volume when the pressure is P,
The volume of the gas becomes

$\frac{1}{2}$ V when the pressure is . . 2 P
$\frac{1}{3}$ V   ,,   ,, . . 3 P
$\frac{1}{4}$ V   ,,   ,, . . 4 P
$\frac{1}{n}$ V   ,,   ,, . . n P

Also

2 V   ,,   ,, . . $\frac{1}{2}$ P
3 V   ,,   ,, . . $\frac{1}{3}$ P
n V   ,,   ,, . . $\frac{1}{n}$ P

Hence the pressure which is produced by the *elastic force* or *tension* of a gas is proportional to its density.

So that if *pressure* increases,
density increases,
and *volume* diminishes.
Also, if *volume* increases,
pressure and *density* diminish.

Law of Boyle.

Boyle's law is not absolutely obeyed by any known gas; but hydrogen and the other gases which are liquefiable with great difficulty conform to the law very nearly, and thus present the nearest approach to the condition of a perfect gas with which we are acquainted. With change of pressure the easily liquefiable gases and vapours increase or decrease in volume to a greater extent than more permanent gases.

The pressure and density of atmospheric air, and of gases which are in communication with it, are estimated by the aid of the barometer. This instrument, in its simplest form, consists of a straight glass tube, somewhat less than a meter long, and closed at one end. The tube is filled with pure mercury, free from air, and then inverted with the open end beneath the surface of pure mercury. The liquid metal then falls from the closed extremity, leaving a space which is generally referred to as the 'Torricellian vacuum.' It contains nothing but mercurial vapour. It is usual to consider that the atmosphere possesses its average and normal density when, at the sea level and at the temperature of 0° C., the column of mercury sustained by the atmospheric pressure is 760 millimetres (or 29·92 inches) high, measuring from the surface of the mercury in the reservoir to the surface of the mercury within the tube. This amount of pressure is often spoken of as one *atmosphere*.

In accordance with the law of Boyle, the volume of a gas under altered barometric pressure can be calculated by the formula

$$\frac{V}{V_1} = \frac{P_1}{P},$$

in which V is the given volume under pressure P, and V_1 is the new volume when the pressure is altered to P_1. So that

$$V_1 = \frac{V}{P_1} \times P.$$

Example.—100 volumes of air are measured off when

the barometric pressure is 740 mm.; what will be the volume of the same air when the barometer stands at 760 mm.?

Here $V = 100$, $P = 740$, $P_1 = 760$.

Then $V_1 = \dfrac{100 \times 740}{760} = 97\cdot 3$ vols.

Answer.

LAW OF GAY-LUSSAC AND CHARLES.

Air expands by $\dfrac{1}{273}$ of its volume at $0°$ for every increase in temperature of $1°$ C.

Thus 273 volumes of air at	.	.	$0°$
become 274	,,	,,	.	.	$1°$
275	,,	,,	.	.	$2°$
276	,,	,,	.	.	$3°$
$273+t$	,,	,,	.	.	$t°$
Also, 273	,,	,,	.	.	$0°$
become 272	,,	,,	.	.	$-1°$
271	,,	,,	.	.	$-2°$
270	,,	,,	.	.	$-3°$
$273-t$	,,	,,	.	.	$-t°$

And generally $273+t$ at $t°$ become $273 + T$ at $T°$.

This fraction $\dfrac{1}{273}$, or $\cdot 003665$, is called the coefficient of expansion, and represents almost exactly the increment or decrement which occurs in a measured volume of air or other permanent gas for every change of temperature of one degree centigrade, provided the pressure remains unchanged. The coefficient of all gases is *very nearly* coincident with that of air, and for chemical purposes may, without inconvenience, be assumed to be the same. Strictly speaking, however, every gas has a coefficient of its own, which, in the case of the more easily liquefiable gases, is perceptibly

Expansion of Gases by Heat. 49

greater than the number given above, as may be seen by the following table:—

Coefficients of expansion for 1° C.

Air	·003665
Hydrogen	·003667
Carbonic oxide	·003667
Nitrogen	·003668
Nitrous oxide	·003676
Carbonic anhydride	·003688
Cyanogen	·003829
Sulphurous anhydride	·003845

It seems not unreasonable to suppose that such differences are due in part to the fact that in vapours and in the more easily liquefiable gases the influence of cohesion is not altogether annulled. It is conceivable that a vapour may consist of molecules which, unlike the independent molecules of perfect gases, may be connected together into companies moving about much in the same way as individual molecules, but less rapidly.

Examples.—A certain mass of air measures 100 cubic centimetres at 0°; to find its volume at 10°.

273 vols. of a gas at 0° become 273 + t vols. when the temp. is t° C.

In this case t = 10.
Then 273 c.c. at 0° become 283 c.c. at 10°,

1 „   „ becomes $\frac{283}{273}$ c.c. at 10°,

and 100 „   „   „ $\frac{283 \times 100}{273}$ c.c. at 10°

Ans. 103·66 c.c.

Or employing the decimal equivalent to $\frac{1}{273}$, let V_1 be the required volume and V the vol. given.

Then $V_1 = V(1 + \cdot 00366\, t)$
$= 100 (1 + \cdot 00366 \times 10)$
$= 103 \cdot 66$ *Ans.*

300 c.c. of air at 20°; find the volume at 0°.

273 + 20 at 20° become 273 at 0°

1 ,, becomes $\dfrac{273}{293}$ at 0°.

And 300 ,, ,, $\dfrac{273 \times 300}{293}$ at 0°.

279·5 *Ans.*

Or, using the decimal coefficient, we say V_0, a certain volume of air at 0°, becomes 300 c.c. at 20°, or

$$V_0 (1 + \cdot 00366 \times 20) = 300.$$

Whence $V_0 = \dfrac{300}{1 + \cdot 00366 \times 20} = 279\cdot 5$ c.c. *Ans.*

500 c.c. of air at 10°; find the volume at − 10°.

In this and all similar problems it is to be remembered that the coefficient of expansion is a fraction of the volume which the gas occupies at 0°, not at any other temperature.

273 + 10 c.c. of air at 10°
measure
273 − 10 c.c. at − 10°.

Therefore 500 c.c. of air at 10°

measure $\dfrac{263 \times 500}{283}$, or 464·6 c.c. at − 10°.

Or let V_{10}, V_0, and V_{-10} be the volumes at the temperatures 10°, 0°, and − 10° respectively; then

$V_{-10} = V_0 (1 - \cdot 00366 \times 10) = \dfrac{V_{10}}{1 + \cdot 00366 \times 10} \times (1 - \cdot 00366 \times 10)$
$= 464\cdot 6$ c.c. *Ans.*

Air or any other permanent gas diminishes by $\dfrac{1}{273}$ of its volume for every degree of temperature travelling down the scale. If the same relations of volume to temperature were maintained, it is obvious that at — 273° the volume would be *nil*, and the gas cease to exist. Such a temperature has, however, never been attained; and if ever such a degree of cold were reached, there can be no doubt that a gas exposed to it would liquefy, or that some change would occur whereby the gas would be released from obedience to the ordinary law.

Notwithstanding, however, that such a condition of things is practically beyond the reach of experiment, this consideration is important as furnishing the basis of an absolute scale of temperature. Calling $-273°$ C. the zero point, we represent absolute temperatures by adding 273 to the number of degrees upon the ordinary Centigrade scale. From what has already been stated regarding the expansion of gases, it follows that, pressure being constant, the volume of a mass of gas varies directly as the absolute temperature. This statement is sometimes referred to as the law of Charles, to whom we owe the discovery[*] of the equal expansibility of the principal gases by heat.

LAW OF AVOGADRO.

It has been shown in the foregoing paragraphs that all gases, when under conditions sufficiently remote from those which induce their liquefaction, are affected in the same manner and to the same extent by changes of pressure and of temperature. Differences of density, of chemical composition, or of chemical properties do not affect the generality of this statement. The volumes of heavy oxygen and light hydrogen, of simple nitrogen and compound marsh gas, increase and decrease according to the same law. It is impossible to avoid the inference from these facts that these gases, so different chemically, must be physically con-

---

[*] Towards the end of the last century.

stituted alike. If now we admit the hypothesis that gases, like other bodies, are made up of small independent masses called molecules, and that heat causes these molecules to separate from one another, whilst cold or pressure causes them to approach, we are led to the assumption that in equal volumes of different gases [1] there must exist the same number of molecules.

This statement, originally enunciated by an Italian physicist, Avogadro, in 1811, may now be regarded as a well-established truth.

But, like every other part of the molecular theory, this law owes its recognition by physicists and chemists, not to any direct proof that can be adduced from experimental sources in support of such hypothesis, but to the fact that nearly all observed chemical phenomena do not only harmonise with such views but find in them complete and satisfactory explanation.

Admitting the law of Avogadro, we see at once why gases are equally expanded by heat, why they are equally contracted by cold and pressure, and why they combine together, according to the discovery of Gay-Lussac, in simple proportions by volume.

In a later chapter will be shown some of the consequences which follow upon an application of this law, and the important progress of chemical theory which has resulted from its adoption.

[1] Under the same circumstances of temperature and pressure.

CHAPTER VIII.

SPECTRA—EMISSION SPECTRA.

ALL bodies when heated to a sufficiently high temperature emit light; those which are densest being, as a rule, the most intensely luminous.

When this light is examined by a prism the image formed by the refracted and dispersed rays appears in the form of a coloured band, which is called the *spectrum*. The apparatus employed for the purpose of observing the spectra of different kinds of light is called a *spectroscope*. The details of its construction are described in nearly all works on physics. Suffice it, therefore, to say that the light under examination is allowed to pass first through a fine slit in a metallic plate so arranged that the slit is parallel to the edges of the prism. The rays are then rendered parallel, by means of a pair of lenses placed in a tube which is fixed at the angle of minimum deviation with the first face of the prism. After passing through the prism the light is viewed through a telescope, which gives a magnified image of the spectrum.

The spectrum of the light emitted by solids, liquids, and very dense gases is usually found to be a continuous one; that is to say, the simple colours—red, orange, yellow, green, blue, indigo, and violet—which are its components merge one into the other gradually, and are not separated by dark intervals. The spectra furnished by ordinary gases or vapours, when ignited, consist, on the contrary, of bright lines, which are so many images of the slit of the spectroscope refracted in different degrees so as to be separated from one another. In the spaces between these bright bands there is in general no light. In some cases, however, a continuous spectrum is more or less distinctly visible.

These differences are shown as far as possible without the aid of colour in the following diagrams. The first represents the appearance which is presented by the spectrum of an ignited solid, such as lime, or of a flame such as that of a candle, which is supposed to contain either solid matters or very dense vapours. The second is the spectrum of heated hydrogen gas.

CONTINUOUS SPECTRUM.

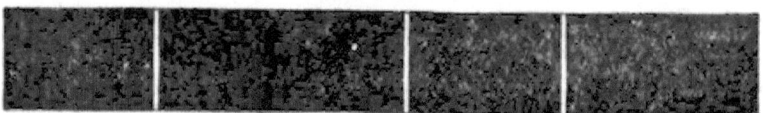

BRIGHT LINE SPECTRUM.

Salts of the alkalis, of the alkaline earths, of copper and many other metals, have long been known to give light having peculiar and characteristic colours when heated in the blow-pipe flame or in the non-luminous flame of the Bunsen burner. Thus, sodium salts give yellow light; potassium, violet; barium, green; calcium, orange-red; and lithium and strontium, crimson flames.

When flames coloured by the vapours of these salts are viewed with the spectroscope, they are found to exhibit a character similar to that of the gas hydrogen. That is to say, these spectra are made up of bright lines which are separated from each other by dark intervals. The appearance shown in each case is indicated by the diagrams on next page.

When the same prism is employed these lines always occupy the same relative positions, and the lines produced by any one substance are not changed in position, or breadth, or intensity in the presence of another substance which, when ignited, gives out a different kind of light.

Line Spectra. 55

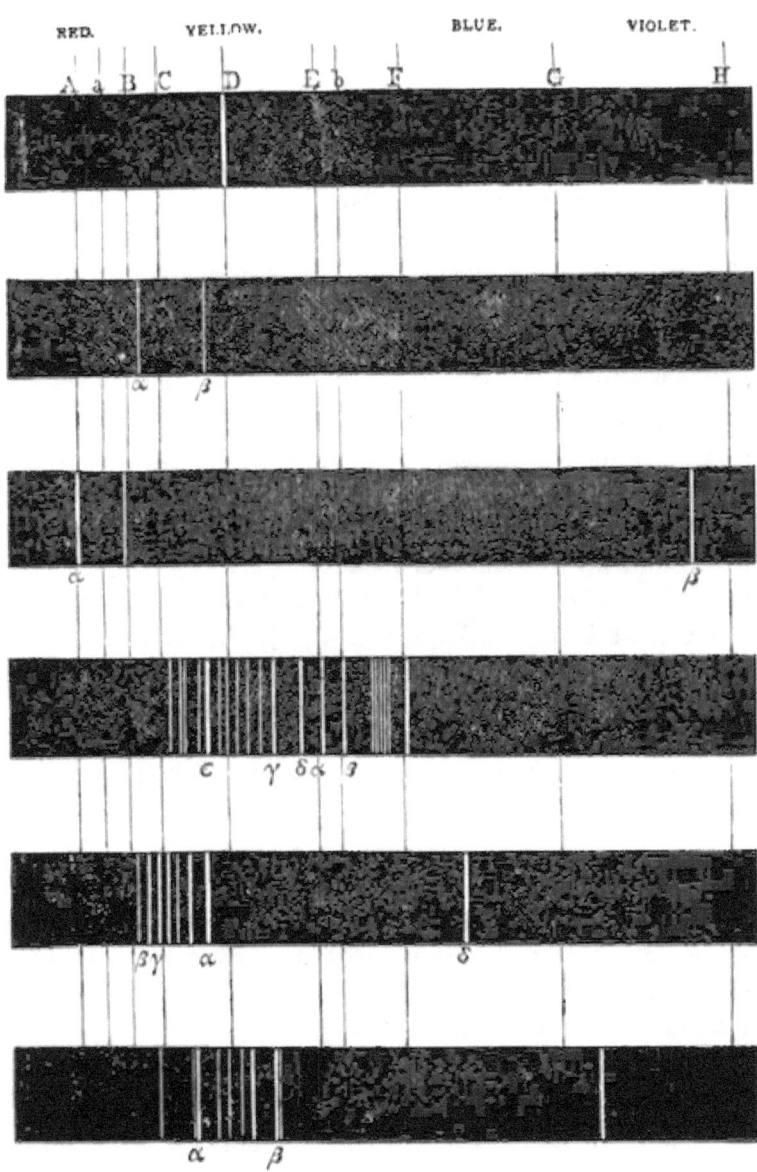

Thus, if a flame coloured by sodium or one of its salts is viewed through the spectroscope, the bright yellow double band characteristic of sodium is alone visible.

Another flame coloured by some compound of potassium gives only the red and deep blue lines peculiar to the ignited vapour of that metal. A third flame, into which is introduced a mixture of potassium and sodium salts, appears yellow to the unaided eye, if the proportion of sodium present is more than infinitesimal. But after passing through the prism of the spectroscope this yellow light is resolved into red and indigo potassium bands and the yellow sodium lines occupying exactly the same position as when observed separately. These facts constitute the basis of the method of spectral analysis.

It is only necessary to bring before the slit of the spectroscope the incandescent vapour of a metal or other substance which it is desired to examine. The position and number of the bright lines visible through the telescope are at once an indication of the nature of the substance. The volatilisation of solid bodies and the necessary ignition of the vapour is effected most generally by the aid of the Bunsen gas-flame; but when this is incompetent to produce a temperature sufficiently high, the oxy-hydrogen blow-pipe or the electric arc may be employed. The transmission of sparks from an induction coil between terminals to which metals can be attached is a convenient method for obtaining the spectra of such substances, as well as of gases such as nitrogen, hydrogen, and carbon dioxide, through which the sparks can be passed.

Another method of obtaining the spectra of gases consists in enclosing them in glass tubes into the walls of which metallic wires are sealed, which serve as poles between which the discharge of an induction coil can pass. On exhausting the tubes by the air-pump so as to remove the greater part, but not the whole, of the enclosed gas, the discharge occurs much more readily than in the same gas under atmospheric pressure, and the attenuated gas becomes brilliantly luminous, and the light can be readily examined by the spectroscope.

The delicacy of the method of spectral observation is very great, far surpassing that of the most exact and sensitive of chemical tests, and by its aid the presence of exceedingly minute quantities of various elements can be detected with certainty and ease. Sodium and lithium, for example, are elements which, even in excessively small quantity, are capable of giving very easily recognisable spectra ; and, accordingly, their presence has been discovered in many substances in which it was formerly unsuspected. Sodium salts are, indeed, almost universally diffused in water, in the mineral constituents of the soil, in the tissues of plants and animals, and in the dust suspended in the atmosphere.

Spectroscopic analysis has also led to the discovery of several elements previously unknown, and existing as unrecognised impurities in various substances. These newly-discovered elements are all metals. Their names are given below :—

NAME.	ORIGINAL SOURCE.	DISCOVERER.	DATE.
Rubidium	Dürckheim Mineral Water	Bunsen	1859
Cæsium		,,	,,
Thallium	Seleniferous pyrites	Crookes	1857
Indium	Freiberg blende	Reich & Richter	1863
Gallium	Zinc blende from Pyrenees	Lecoq de Boisbaudran	1875

It will be seen by reference to the diagrams that the spectra of the metals of the alkalis and of the alkaline earths are comparatively simple, consisting, in most cases, of a small number of lines, which are often widely separated and easily recognised. Several other metallic elements, under the same conditions, yield equally simple spectra : that of thallium, for example, consisting of a single green line, whilst the spectrum of indium exhibits one line in the blue and another in the indigo. The spectra of the heavy and less volatile metals are, however, in general, much more complex; the spectrum of iron, for instance, showing upwards of four hundred and fifty lines, many of them crowded together in the green.

When a compound, such as the chloride of a metal like sodium, is heated in the Bunsen flame, it gives out light which is very generally identical with the light obtained from the incandescent vapour of the metal itself. Sometimes, however, this is not found to be the case, and the spectra of certain compounds is different from that of either of the constituents, taken separately, and is, in general, more complex.

The accompanying diagram exhibits a comparison of the spectra obtained by the introduction of solid calcium

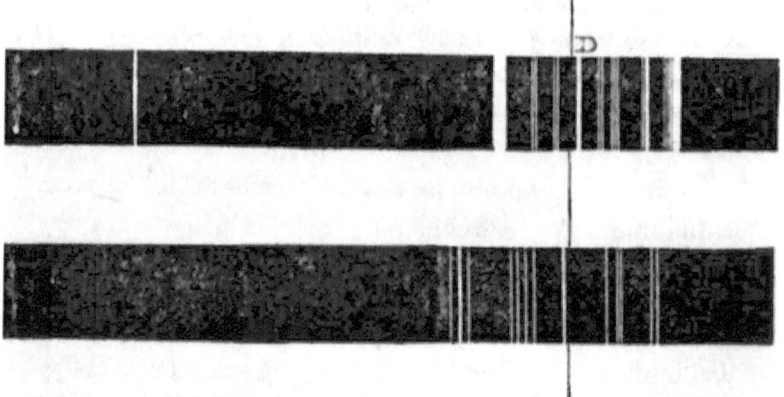

chloride into the non-luminous gas-flame, and when the electric spark is passed over the same compound.

The higher temperature in the second case causes a decomposition of the compound, and the observed lines, which are changed both in number and in refrangibility, are attributed to the glowing vapour of the metal calcium itself.

The position of such lines is not altered at still higher temperatures, though not unfrequently new lines make their appearance.

An interesting method of applying the spectroscope to the examination of solid bodies has been much employed in investigations into the nature of the rare earths which contain yttrium and its allies. When the spark from a good induction coil traverses a tube provided with a flat aluminium pole at each end, and the air is then removed by the Sprengel mercury-pump, the luminosity which the gas exhibits in the early stages of the exhaustion gradually disappears, and a dark space which at first surrounds the negative pole gradually extends throughout the tube. When the luminous cloud showing the presence of residual gas has almost disappeared, the molecular discharge from the negative pole begins to excite phosphorescence on the glass where it strikes the side of the tube, and in any substance enclosed within the tube. This occurs with the majority of substances only when the pressure of the gas within has been reduced to about one-millionth of an atmosphere. The emitted light yields in most cases a faint continuous spectrum; but in some cases, especially the sulphates of the earths referred to above, a spectrum is observed consisting of bright lines.

ABSORPTION SPECTRA.

When a transparent coloured medium, such as a piece of glass or a coloured liquid, is brought into the path of a ray of light before it enters the spectroscope, certain portions of the spectrum disappear. The bands of darkness which are thus produced are due to the interception of certain portions of the light by the coloured glass or solution, or vapour, as the case may be. The resulting spectrum is called an *absorption spectrum*. It presents characteristics which in many cases are as decided as those of the emission spectrum, and the observation of absorption spectra may occasionally be turned to practical account in the recognition of the colouring matter of blood, of wines, and many other substances, as well as of certain coloured gases and vapours.

When the absorbing medium is the vapour of an element

or of a compound which is volatile without decomposition, the dark lines of absorption occupy the same position as the bright lines in the spectrum of the light produced by the ignition of the same vapour. Thus, if a beam of light from a lamp is allowed to traverse a sufficiently thick stratum of sodium vapour, two dark lines close together make their appearance in the yellow ; and if such a spectrum is viewed side by side with the ordinary spectrum of the monochromatic sodium light, the position of the dark absorption band is seen to coincide with that of the bright yellow lines. Such phenomena are explained by the law that gases and vapours are capable of absorbing and stopping the same rays of light which they emit at higher temperatures, when in the state of ignition or incandescence.

This law has been experimentally verified by the examination not only of the absorption spectrum of sodium, but of other elements which give more complex spectra. The facts thus established have been employed in the solution of the very interesting problem presented by the light which reaches us from the sun and other heavenly bodies. Sunlight examined by a spectroscope exhibits a spectrum from which none of the primary colours are absent, but which is traversed by a very large number of fine black lines, many of which were discovered so long ago as 1802 by Wollaston, and examined in 1814 by a German optician, Fraunhofer. They are usually known as Fraunhofer's lines.

In the figure, the position of a few only of the most prominent is indicated, and they are distinguished by the letters used originally by Fraunhofer.

These dark spaces are now known to be absorption bands. It has been found that their positions coincide with the lines of the spectra of many terrestrial elements; and the coincidences which have been observed are so absolute and so numerous as to lead to the inevitable conclusion that they are produced by the vapours of such bodies existing in the gaseous envelope of the sun.

The eye enables us to recognise only a part of the radiation from a luminous object, namely, those rays only which correspond to a rate of vibration lying within certain comparatively narrow limits, and which possess a particular range of refrangibility. But accompanying these are others, on the one hand of greater wave-length, and consequently smaller refrangibility, which, in passing through a prism, are less bent than the extreme red of the visible portion; and on the other hand there are rays of shorter wave-length and greater refrangibility than those which are found at the extreme visible portion of the violet. These rays, though they give to the eye no impression of light, are capable of producing both heating and chemical effects. They are not, however, equally capable of producing such effects.

A sensitive thermometer exposed successively to different portions of the visible spectrum will indicate that the heating effect is produced chiefly towards the red end, and even that the maximum is attained amongst the rays which are less refrangible than the extreme end of the visible red, and the spectrum of which lies beyond in darkness. The yellow and red luminous rays and the dark heating rays possess, however, little or no chemical action so far as ordinary silver salts are concerned, and accordingly photographers are accustomed to manipulate their sensitive plates in rooms illuminated by yellow or red light. The chemically active rays are confined almost entirely to the violet end of the spectrum; the exact position of the most active being found, however, to depend to a certain extent upon the nature of the substance submitted to their influence. Thus, silver salts are blackened, and hydrogen and chlorine are caused to combine most rapidly, when exposed to that part of the violet end situated between the lines G and H of the solar spectrum, though more or less chemical activity is manifested some distance beyond the visible violet on the one side and as far as the middle of the green on the other ; all perceptible action ceasing in the yellow or most luminous portion of the visible spectrum.

These relations are exhibited by the three curves in the following diagram :—

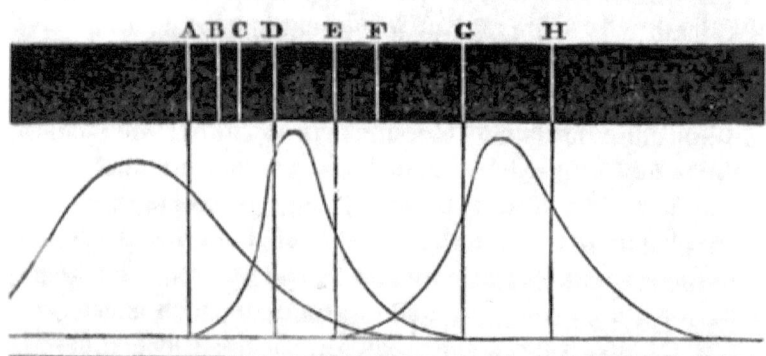

The summit of each curve indicates the position of the maxima of heating, luminous, and ordinary photographic effects in the spectrum of sunlight, the vertical lines representing the positions of the chief dark lines in the solar spectrum.

It would not be correct, however, to assume that rays in other parts of the spectrum have no chemical activity, inasmuch as this property depends not only upon the colour and refrangibility of the ray, but also upon the nature of the substance upon which it falls. The chemical effects observed in photographic plates are due to absorption of the rays by the sensitive film; and when the composition of this film is altered, or addition made of some appropriate colouring matter, a photographic or chemical effect may be obtained in any part of the spectrum from the region beyond the extreme visible violet to the dark region beyond the red. These curves are represented as overlapping one another, but it must not be supposed from this that there are three distinct sets of rays in the spectrum. In all probability the rays are of the same kind from end to end, but differ in wave-length and rapidity of vibration. The least refrangible rays are capable of producing the effects of heat, but are not capable of exciting the sensations of vision; whilst the vibrations which communicate the sensation of violet to the eye are also capable of effecting chemical change, though their heating power is exceedingly small.

Modern researches having shown that invisible rays at both extremities of the spectrum are capable of producing a photographic image, when allowed to fall upon a plate coated with suitable material, this discovery has rendered it possible to observe the absorption-spectra of various transparent media beyond the limits of the visible portion. The most interesting results have been obtained with liquids which are usually regarded as colourless, such as water, alcohol, and various hydrocarbons. These liquids for the most part absorb and stop certain rays both in the visible portion of the spectrum and in the invisible regions at the two extremities, and by the position of the dark bands or lines thus observed they may be readily identified. The only liquids which have at present been observed to give no absorption bands either in the infra-red region or in the visible spectrum are carbon tetrachloride and carbon disulphide. All those which contain hydrogen have been found to give lines and bands of absorption.

So far as concerns the highly refrangible rays in the ultra-violet, Professor Hartley concludes that though a general absorption may be observed in the case of many alcohols, acids, and hydrocarbons, it is only benzene and compounds which contain six-carbon atoms united together in the same manner as in benzene that give definite absorption bands. (See formula of benzene, p. 319.)

LUMINOSITY OF FLAME.

The cause of the luminosity of common candle and gas flames has long been a subject of interest, and two chief hypotheses have been framed with the object of explaining it. Though neither of these hypotheses alone is capable of furnishing a complete explanation of every case, yet, taken together, they serve to account satisfactorily for the phenomena which are usually observed.

According to the earlier hypothesis, proposed in 1817 by Sir H. Davy, the luminosity of flame is attributed to the existence in the flame of particles of solid matter, which, being heated to a high temperature by the burning gases, emit light. These solid matters, supposed to exist in the flame, are deposited in the form of soot when a cold body is plunged into the flame. Such facts as the following were adduced by Davy in support of his hypothesis.

The flame of hydrogen or of alcohol, burning in the usual way, emits only a very feeble light; but the introduction of solid matter, such as powdered charcoal, oxide of zinc, or dust of any kind, whether combustible or not, serves to render such a flame luminous.

A bright flame is also produced by the combustion, in air or oxygen, of substances which, like metallic zinc, form solid products of combustion. On the other hand, sulphur, hydrogen, and carbonic oxide, which, in burning, yield entirely gaseous products, give out light very sparingly.

The spectroscope has also shown us that ordinary luminous flames yield a continuous spectrum of the same character as that usually produced by incandescent solids.

The production of solid particles in a candle-flame, or in the flame of coal-gas, or other hydrocarbon, is supposed to be due to the selective power of the oxygen of the air, in virtue of which it unites preferably with the hydrogen, leaving a part of the carbon in a solid form to be consumed as it reaches the higher parts of the flame.

A candle flame burning steadily exhibits the form of a long cone with the apex pointing upwards. Other flames taper upwards in the same way. This form is, of course, due to the strong upward current produced in the air which immediately surrounds the flame, in consequence of the heat. A close examination of the flame of a candle will show that it consists of several parts At the base is a faintly luminous stratum of a blue colour. Higher up the luminosity increases, the most intense light proceeding from those parts which are near the middle; whilst the whole structure is surrounded by a transparent and almost invisible, but very hot, envelope of ignited gas. It is easy to show by many simple experiments that such a flame is only a shell of ignited gas, the process of combustion occurring only on the outside, the interior being filled with comparatively cool gas or vapour. Thus, if a sheet of paper is held for a moment horizontally across the flame it will receive a deposit of soot in the form of a ring. A slip of wood or a wire passed through the flame becomes ignited only at the two points where it cuts the outer portions of the flame, the part in contact with the interior remaining dark.

The luminosity of such a flame is diminished if a sufficient quantity of air or oxygen is thrown into it, and at the same time it loses the power of depositing soot upon any cold object held in it. The flames of the blowpipe and the Bunsen burner are produced in this way, and present this character.

Some experiments made a few years ago by Dr. Frankland indicate that the ignition of solid particles of sooty matters in hydrocarbon flames may occasionally be the cause of the light emitted by such flames, but that in some cases at least the effect is not wholly attributable to this circumstance. Dr. Frankland has observed that when hydrogen is burnt in oxygen under great pressure the light of the flame, usually so pale, is increased to such an extent as to be capable of illuminating a page of print so that it can be read at some distance.

The spectrum of hydrogen burning under pressure exhibits the three bands characteristic of hydrogen, but much broader and more or less nebulous at the edges, so that an approach to a continuous spectrum similar to that obtained from solid bodies is the result. A similar effect is produced by burning carbonic oxide. So that the argument founded upon the nature of the spectrum of an ordinary luminous flame has less weight than might otherwise be supposed.

Moreover, in burning a series of substances, all of which yield volatile products of combustion, it is found that many of them are capable of emitting a very vivid light, quite equal in brilliancy to the light produced by the ignition of many substances which are solid and not capable of vapourisation. Thus, the combustion of arsenic in oxygen is attended by the emission of a very brilliant white light, although all the substances present—the arsenic, the oxygen, and the arsenious oxide which is formed—are, at the temperature of ignition, entirely in a state of vapour.

In such cases, therefore, the light cannot be attributed to glowing solid matter. It has been observed that the brilliancy of the light emitted by the combustion of such substances is nearly proportional to the density of the ignited vapours existing in the flame, provided that in each case the temperature is sufficiently high.

In all probability, then, the luminosity of burning gas or tallow is due to the ignition of vaporous, but very dense, hydrocarbons, and is not to be ascribed to the presence of solid particles of carbon. The fact that ordinary flames are transparent is also difficult to reconcile with the latter hypothesis.

EXERCISES ON SECTION I.

1. Water is shaken up with a large volume of oxygen gas under a constant pressure of 765 mm. What volume of the gas will be contained in 10 c.c. of the solution?

2. Water is exposed to an atmosphere consisting of 21 vols. of oxygen, with 79 volumes of nitrogen.

Temp. $0°$; pressure, 760 mm.
Coeffs. of sol., $N = .02$, $O = .04$.

Calculate (a) the total volume of gas dissolved by 52·5 c.c. of water, and (b) the percentage composition of the gas.

3. Soda-water is charged under a pressure of 2·3 atmospheres. Calculate the volume of carbonic anhydride contained in 300 cubic centimetres of such water.

An atmosphere = 760 mm. barom.
Coeff. of sol. for carbonic anhydride, 1·7967.

4. Sulphur dioxide is passed into water as long as it is absorbed. If the barometer stands at 745 mm., calculate the volume of gas contained in half a litre of the solution.

Coeff. for sulphur dioxide, 79·789.

5. Water is shaken up with its own bulk of a mixture of 1 volume of oxygen with 3 vols. of nitrogen. Supposing the temp. and pressure to remain normal and constant throughout the experiment, calculate the composition of the residual air.

Coeff. of oxygen, ·04114.
Coeff. of nitrogen, ·02035.

Diffusion of Gases.

6. The specific gravity of chlorine is 35·5 ($H = 1$). Compare its velocity of diffusion with that of hydrogen.

7. Specific gravity of ozone, 24; of carbonic anhydride, 22. Compare their velocities of diffusion with each other and with that of H (sp. gr. 1).

8. The rate of diffusion of a gas is observed to be ·81 when that of air is 1. Find its density.

9. Oxygen and hydrogen are separated by a porous plate, and 3·83 cubic centimetres of hydrogen pass through the plate in a second.

What volume of oxygen passes during the same time in the opposite direction?

10. In the last question suppose the original volume of the oxygen to have been 20 c.c.: what will be the composition of the mixture formed in its place after three seconds, assuming the apparatus so arranged that no change of pressure occurs?

Corrections of Gas-volumes for Changes of Pressure and Temperature.

11. 100 c.c. of air when bar. = 750 mm. Find the volume when bar. = 790 mm.

12. 250 c.c. of air when bar. = 765 mm. Find the volume when bar. = 745 mm.

13. What pressure in atmospheres would be required to make the density of hydrogen (sp. gr. ·0693) equal to that of air?

14. Calculate the atmospheric pressure per square centimetre when the barometer stands at 760 mm.

Weight of 1 c.c. of mercury, 13·596 grams.

15. Find the atmospheric pressure per square decimetre when the barometer stands at 750 mm.

16. What change of atmospheric pressure will be denoted by a change of 12 mm. in the barometric column?

17. The weight of one litre of hydrogen at 0° and 760 mm. is ·0896 gram or 1 crith. Calculate the weight of one litre of hydrogen measured off under a pressure of 1400 mm.

18. Find the weight of 1 litre of nitrogen (sp. gr. 14); of 10 litres of carbonic anhydride (sp. gr. 22); of 250 c.c. of oxygen (sp. gr. 16).

19. A mass of air at 0° measures 100° c.c. What volume will it occupy at 20°; at 15°·5; at 100°?

20. A certain quantity of air is measured at 75°. What volume will it have at 0°?

21. 1000 c.c. of a gas at 12°·5. What volume at 75°?

22. 500 c.c. of a gas at 10°. What volume at 40°?

23. 300 c.c. of a gas at 25°. What will its volume be at − 10°?

24. 75 c.c. of nitrogen measured at 50°. What volume at − 35°?

25. 1500 c.c. of hydrogen measured at 20°. At what temperature will it measure 1000 c.c.?

26. Five degrees centigrade correspond with nine degrees on the Fahrenheit scale. Find the coefficient of expansion of gases for 1° F.

27. 150 c.c. of nitrogen are measured at 10°, and under a pressure

of 500 mm. of mercury. What will the volume become at 16°·4 when the pressure is 540 mm.?

28. A quantity of nitrogen confined in a tube standing over mercury in a mercurial trough measures 75·5 c.c.; temp. 15°; bar. 742 mm.; surface of mercury inside the tube above surface of mercury in the trough 122 mm. Find the volume which the gas would occupy at normal temperature and pressure.

29. 1000 cubic feet of gas are put into a balloon of 1250 cubic feet capacity; temp. 18°; bar. 765 mm. After ascending a certain height it is found to be fully distended. What is the atmospheric pressure, temperature being 8°?

30. A certain balloon is just capable of holding 10 grams of hydrogen under standard conditions: what is its capacity? How much larger must it be made if it is required to sustain a diminished atmospheric pressure equal to 650 mm. bar.?

SECTION II.

CHAPTER IX.

ELEMENTS AND COMPOUNDS.

WHEN water is exposed to a very high temperature or made the vehicle of an electric current, it disappears and is replaced by an equal weight of a mixture of two gases, hydrogen and oxygen. These two gases will, under certain conditions, again give rise to water and to exactly the same amount of water as at first.

Water then is said to be composed of oxygen and hydrogen. It is worth noting, however, that, strictly speaking, this can only mean that in proportion as the water is destroyed or ceases to exist, the gases make their appearance, and *vice versa*, for in water we have no resemblance to hydrogen or oxygen, neither can we detect either of those bodies in water except by this process of so-called *decomposition*.[1]

Now, if the hydrogen thus obtained from water is submitted to a repetition of the same kind of treatment or to any other that may suggest itself, it refuses utterly to yield up anything that is not hydrogen. In other words, it cannot be decomposed. We find then that certain bodies,[2] such

---

[1] 'Cavendish and Watt both discovered the composition of water. Cavendish established the facts; Watt, the idea. Cavendish says, "From inflammable air and dephlogisticated air water is produced." Watt says, "Water consists or is composed of inflammable air and dephlogisticated air." Between these forms of expression there is a wide distinction.'—*Liebig's Letters on Chemistry*, p. 58.

[2] It must be understood that bodies of definite characters, and not mere mixtures, are here referred to. (See Chap. XII.)

as water, may be resolved into two or more different kinds of matter, and these are called *compounds*; whilst others like hydrogen cannot be split up in this manner by any means with which we are at present acquainted, and are regarded as *elements*.

It must be understood that the word 'element' is only provisional, and is merely intended to imply that in the present state of knowledge the bodies so called have resisted all attempts to decompose them.

Concerning the primal elements of nature nothing whatever is known. Some speculations concerning the constitution of the chemical 'elements' will be referred to later on, but it will be sufficient in this place to indicate the general nature of the investigations that have been made in regard to one or two of them.

Chlorine was discovered in 1774 by Scheele, who, in the language of the then prevalent theory, called it 'dephlogisticated muriatic acid.' This was equivalent to saying that it was muriatic acid deprived of its inflammable principle, hydrogen. After the discovery of the fact that many of the common acids contain oxygen, Berthollet, acting under the belief that it contained oxygen, gave it the name 'oxymuriatic acid.' Davy proved, in 1810, that it contained neither hydrogen nor oxygen, and gave it the name by which it has since been known and ranked with the rest of the elements.

Davy's chief experiments were as follows :—

The dry gas mixed with dry hydrogen unites with it without contraction and without the production of any water. Charcoal, previously freed from hydrogen and moisture, undergoes no change when intensely ignited in the dry gas by the voltaic current even for several hours, and no oxide of carbon is formed. Potassium, sodium, tin, copper, and other metals when heated in it combine with the whole of the gas, leaving no gaseous residue and forming no oxide of the metal. Phosphorus and sulphur, both highly oxidisable substances, unite with chlorine but set free from it no muriatic acid. Chlorine mixed with dry ammonia produces only nitrogen and dry sal-ammoniac, and no water is formed. Electric sparks passed for a long time through chlorine gas produce no change in it. When chlorine acts upon a heated metallic oxide the amount of oxygen expelled is equal to that which is contained in the oxide and no more. So much for direct experimental evidence. The discovery of iodine, in 1811, and of bromine, in 1826, confirmed the position of chlorine among the elements, for if oxygen be a constituent of one of these bodies it must be present in them all. Further knowledge since obtained, the accurate determination of their atomic weights and the place they occupy in the system of classification which is based upon the numerical values of the atomic weights, establishes the claim of these three substances to be regarded as a natural group of elements. (See Chap. XXIII.)

The case of the fixed alkalies, potash and soda, is the converse of that of chlorine. Originally supposed to be elementary in consequence of the failure of all attempts to extract from either of them more than one form of matter, they had been suspected by Lavoisier to contain nitrogen, probably by reason of a certain resemblance to the volatile alkali ammonia, which was known to contain both hydrogen and nitrogen. By some of the same school it had been guessed that the alkalies and alkaline earths were compounds of metals with oxygen, in consequence of the resemblance which some of them, such as lime, present to oxides, like litharge, the composition of which was known. In 1807 it was known that the voltaic current would decompose water into oxygen and hydrogen, and that solutions of various salts were resolved into acid, which collected at the positive pole, and alkali, which always passed to the negative, also that many of the metals and their oxides might be separated from their compounds by the same agency. With these facts in view Davy exposed caustic potash to the action of a powerful electric current and obtained oxygen and potassium. Caustic soda was by the same means shown to contain sodium. These new metals were at first supposed to be compounds of hydrogen with potash and soda respectively, but in that case they ought, when burnt in oxygen, to produce the hydrates. That this is not the case was shown by Davy.

As to potassium and sodium their elemental nature seems almost above suspicion, for there is no class of substances possessing so many characters in common as the metals, and if it could be shown that any one of them contained, for example, hydrogen, the same constituent might be expected to occur in all. The results of spectroscopic analysis, however, seem to place this question beyond doubt, for the spectrum of each metal is independent, and shows no sign of containing an element in common with any of the others.

Other examples might be mentioned of substances long regarded as chemical elements ultimately showing their true character as compounds under the influence of some new agent or some improved mode of operating; thus, the substance regarded as metallic vanadium by Berzelius turned out to be an oxide of vanadium when examined long afterwards by Roscoe.

In the case of oxygen it is found that, by the action of electricity and otherwise, it may be converted into another gas, ozone, possessing remarkable characters quite distinct from those of oxygen. Nevertheless, oxygen ranks as an element because it yields in this way only one new body at a time, which by mere application of heat recovers its original properties, and that without loss or gain in weight. It must, therefore, be assumed that the altered properties exhibited under these circumstances result from a temporary rearrangement of its constituent particles. When, as in this case, elementary matter, stuff, or substance is capable of making its appearance in the form of two or more bodies having different properties, these are said to be *allotropic* modifications of the element, and the phenomenon is spoken of as allotropy. (See 'Isomerism,' Chap. XXI.)

About seventy elements are known at the present time, but it is not improbable that a few new substances may be hereafter added to this number. It is not, however, very likely that any hitherto unknown elements will be found to occur in any considerable quantity among the constituents of the earth's crust, for every substance within reach of man has already been subjected to a very close scrutiny by chemists.

It will be seen by reference to the following table that the materials which compose the solid earth, so far as we know it—the ocean, the atmosphere, and the bodies of the living beings which inhabit it—are made up of a few of these elements, the rest occurring in much smaller quantity, in some cases discoverable only by specially delicate methods.

Water consists of . . . . { Hydrogen. Oxygen. }

Air consists chiefly of . . . { Oxygen. Nitrogen. }

Solid earth consists chiefly of
- Silica = { Silicon. Oxygen. }
- Limestone = { Calcium. Magnesium. Carbon. Oxygen. }
- Various silicates forming crystalline rocks or beds of clay { Silicon. Oxygen. Aluminium. Iron. Calcium. Potassium. }

Plants consist chiefly of . . . { Carbon. Hydrogen. Oxygen. }

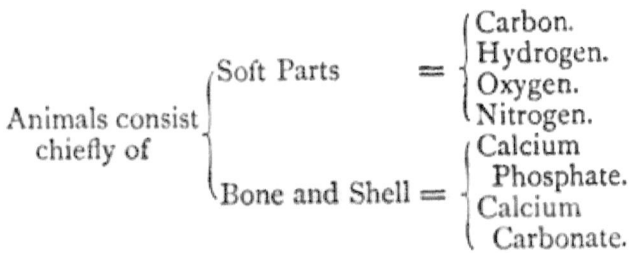

Chemists have found it convenient to adopt a system of symbols and formulæ, whereby to represent the elements and those compounds the composition of which is known.

Thus, to each of the elements is assigned a symbol formed generally of the initial letter of the Latin name of the element. For example, sulphur is represented by the symbol S; selenion, Se; silicon, Si; strontium, Sr; but silver (*argentum*) is Ag, and sodium (*natron*) has the symbol, Na.

But chemical symbols are not merely abbreviations contrived, like shorthand characters, for the purpose of saving trouble in writing the names. Each symbol represents *one atom* of the element for which it stands, and hence expresses a definite weight and volume of the element, which are identical with the proportions by weight and volume in which it enters into chemical combination.

The weight of an atom of hydrogen is less than the weight of an atom of any other element. It is, therefore, convenient to consider the value of the symbol of hydrogen as unity and the values of all other symbols greater than unity. But since nothing is known as to the absolute weight of an atom of hydrogen or of any other element, it should be borne in mind that the atomic weights are in reality ratios, or fractions whose denominator is 1, although they are always written in the form of whole numbers.

In the following list are given the names of all the known elements, together with their symbols and atomic weights, and the student is recommended at once to commit to memory those which are printed in capital letters, leaving the rest to be learnt gradually as occasion may require. In a later chapter an account will be given of the methods by which the numerical values of the atomic weights are determined.

TABLE OF ATOMIC WEIGHTS.

Name of Element	Symbol	Calculated from the most accurate Experiments (H = 1)	Approximate Values for Common Use, O = 16
ALUMINIUM	Al	27·009	27
Antimony	Sb (Stibium)	119·955	120
Arsenic	As	74·918	75
Barium	Ba	136·763	137
Beryllium	Be	9·085	9·1
Bismuth	Bi	207·523	208
Boron	B	10·941	11
BROMINE	Br	79·768	80
Cadmium	Cd	111·835	112
Cæsium	Cs	132·583	133
CALCIUM	Ca	39·990	40
CARBON	C	11·9736	12
Cerium	Ce	140·424	141
CHLORINE	Cl	35·370	35·5
Chromium	Cr	52·009	52·1
Cobalt	Co	58·887	59
COPPER	Cu (Cuprum)	63·173	63·3
Didymium	Di	—	145 ?
Erbium	E	165·891	166
FLUORINE	F	18·984	19
Gallium	Ga	68·854	69
Germanium	Ge	72·32	72
Gold	Au (Aurum)	196·850	197
Holmium	Ho	?	?
HYDROGEN	H	1·0000	1
Indium	In	113·398	114
IODINE	I	127·557	127
Iridium	Ir	192·651	193
IRON	Fe (Ferrum)	55·913	56
Lanthanum	La	138·526	139
LEAD	Pb (Plumbum)	206·471	207
Lithium	Li	7·0073	7
MAGNESIUM	Mg	23·959	24
Manganese	Mn	53·906	54
MERCURY	Hg (Hydrargyrum)	199·712	200
Molybdenum	Mo	95·527	95·7
Nickel	Ni	57·928	58

Symbols and Atomic Weights. 77

TABLE OF ATOMIC WEIGHTS—*continued.*

Name of Element	Symbol	Calculated from the most accurate Experiments ($H = 1$)	Approximate Values for Common Use. $O = 16$
Niobium	Nb	93·812	94
NITROGEN	N	14·0210	14
Osmium	Os	198·494	199
OXYGEN	O	15·9633	16
Palladium	Pd	105·737	106
PHOSPHORUS	P	30·958	31
Platinum	Pt	194·415	195
POTASSIUM	K (Kalium)	39·019	39·1
Rhodium	Ro	104·055	104
Rubidium	Rb	85·251	85·5
Ruthenium	Ru	104·217	104
Samarium	Sm	?	?
Scandium	Sc	43·980	44
Selenion	Se	78·797	79
SILICON	Si	28·195	28
SILVER	Ag (Argentum)	107·675	108
SODIUM	Na (Natrium)	22·998	23
Strontium	Sr	87·374	87·6
SULPHUR	S	31·984	32
Tantalum	Ta	182·144	182
Tellurium	Te	127·960	126 ?
Thallium	Tl	203·715	204
Thorium	Th	233·414	234
Thulium	Tm	?	?
Tin	Sn (Stannum)	117·698	118
Titanium	Ti	48·013	48
Tungsten	W (Wolfram)	183·610	184
Uranium	U	238·482	239
Vanadium	V	51·256	51·4
Ytterbium	Yt	172·761	173
Yttrium	Y	89·816	90
ZINC	Zn	64·9045	65
Zirconium	Zr	89·367	89·6

In order to indicate combination between two elements their symbols are placed side by side, thus, HCl. When a molecule of a compound contains more than one atom of each or either of its constituents, the number of atoms is indicated by a small figure placed below the line.

Thus the formula for water, OH_2, represents one atom of oxygen united with two atoms of hydrogen, and H_3PO_4 means that three atoms of hydrogen, one atom of phosphorus, and four atoms of oxygen are bound together in phosphoric acid.

The formula, taken as a whole, is generally assumed to represent *a molecule* of the compound, though it will hereafter (Chapter XV.) be explained that much uncertainty exists in reference to all substances except such as are vaporisable without chemical decomposition. The relative weight of the molecule is easily found by adding together the weights represented by the several symbols of which it is made up. This weight is called the molecular weight. In order to express two or more molecules a figure is placed at the beginning of the formula, and must be understood to multiply every symbol that follows it. For example, $2OH_2$ represents two molecules of water, each consisting of two atoms of hydrogen with one atom of oxygen; in all, four atoms of hydrogen and two atoms of oxygen. Occasionally brackets have to be introduced when some group of symbols occurs more than once. Aluminium sulphate, for example, has the formula $Al_2(SO_4)_3$, which is thus written for the sake of assimilating the appearance of the formula to that of other sulphates, such as H_2SO_4, $BaSO_4$, $FeSO_4$, &c. Resemblance between them would be less apparent if it were expressed as $Al_2S_3O_{12}$.

Care and a little practice are all that is necessary to avoid confusion in the use of formulæ, and the student is therefore recommended to work out conscientiously all the examples given at the end of the section.

CHAPTER X.

LAWS OF CHEMICAL COMBINATION.

THE extraction of the metals from their ores, the manufacture of alkalies, soap, glass, dyes, and a variety of other useful applications of practical chemistry, were known to man in a more or less practical form from very early times. The alchemists extended the art of chemistry by the discovery of the processes for producing many acids and salts, and by the invention of much useful apparatus. The discovery of oxygen by Priestley, of chlorine by Scheele, the proof of the composition of water supplied by Cavendish, and the overthrow of the phlogistic theory by Lavoisier's explanation of combustion, were all great strides in advance which followed one another in rapid succession. But the foundation of chemistry as an exact science was only laid towards the end of the last century, when the balance began to be used systematically in all chemical investigations. Exact determinations of the relative weights of bodies engaged in various chemical reactions were necessary for the establishment of those laws upon which chemical ideas of the present day are founded. If we start by admitting the molecular constitution of matter, it becomes unnecessary to make any formal statement of the laws which have been found to regulate the distribution of weight or quantity of matter involved in chemical actions. If we assume, for example, that the element oxygen is made up of molecules, all having the same properties, and each composed of two atoms each having a weight sixteen times that of the hydrogen atom, it is obvious that only a definite number of those atoms can take part in forming a chemical compound. And if, as we believe, these atoms are indestructible by chemical action, no fractional part of an atom can enter into such process; and if the number of atoms thus employed be represented as n, the weight of the substance will be 16n.

But since the molecular theory was not always employed by chemists, and even at the present day is not adopted universally, it is desirable still to indicate the facts which have been observed in connection with this question. The four general statements or laws may be expressed as follows:

LAW OF DEFINITE PROPORTIONS.

The proportions in which bodies unite together chemically are definite and constant. In other words, a given chemical compound always consists of the same elements united in the same proportions.

In order to form water, for example, union between hydrogen and oxygen occurs exactly in the proportion of two measures of the former to one measure of the latter. This corresponds with two parts by weight of hydrogen to sixteen parts by weight of oxygen, since oxygen is sixteen times heavier than hydrogen. The employment of any larger quantity of either element would only result in the excess being left uncombined.

LAW OF MULTIPLE PROPORTIONS.

When two substances unite together in several proportions, then if one of them is taken as the unit, the quantity of the other substance in the different compounds varies in a simple manner. Oxygen combines with carbon, forming two oxides of carbon. In the one, three parts by weight of carbon are united to four parts of oxygen, whilst in the other three parts of carbon combine with twice this quantity or eight parts of oxygen. A great number of equally simple cases might be cited, and, doubtless, it was the study of such cases which in the first instance led to the enunciation of the law. In perhaps a still greater number of instances, however, it is by no means easy to trace its application. Among the numerous compounds of carbon with hydrogen,

Multiple and Reciprocal Proportions. 81

for example, are found such relations as the following, which represent the composition of the series of paraffins :—

	Carbon.		Hydrogen.	
Methane	3	parts with	1	part
Ethane	4	,,	1	,,
Propane	9	,,	2	,,
Butane	24	,,	5	,,
Pentane	5	,,	1	,,
Hexane	36	,,	7	,,
Heptane	21	,,	4	,,
Octane	16	,,	3	,,
Nonane	27	,,	5	,,
Decane	60	,,	11	,,
&c.	&c.		&c.	

LAW OF RECIPROCAL PROPORTIONS.

The weights of two different elements, A and B, which combine with a third, C, represent the proportions in which they will themselves unite together if union between them is possible, or they bear some simple relation to those proportions. Thus, $35\frac{1}{2}$ parts of chlorine and 80 parts of bromine combine with 23 parts of sodium. Then, according to the law, when chlorine combines with bromine, $35\frac{1}{2}$ parts of the former are required for every 80 parts of the latter. This may be rendered graphically somewhat in this manner :—

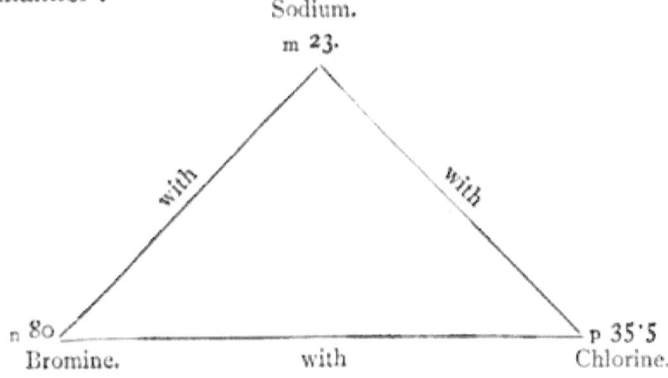

In other words, the combining weight of a body may always be represented as a multiple of the same number whatever state of combination it enters into, or $m\,n$ is the combining weight, where m is a number peculiar to the body and n is some integer.

LAW OF VOLUMES.

When gases combine together they do so in equal volumes, or in volumes which have some simple relation one to another, as 1 to 2, 1 to 3, 2 to 3, and so on. For example, hydrochloric acid is formed by the union of

>1 volume of hydrogen with
>1 volume of chlorine;

water is formed by the union of

>2 volumes of hydrogen with
>1 volume of oxygen;

ammonia by the combination of

>3 volumes of hydrogen with
>1 volume of nitrogen;

nitrous anhydride by the combination of

>2 volumes of nitrogen with
>3 volumes of oxygen, &c.

These facts are recorded in the formulæ

$$HCl \qquad H_2O \qquad H_3N \qquad N_2O_3$$

by which these compounds are commonly represented.

It is important also to remember that whatever the volume of the vaporous elements before combination, the bulk of the resulting compound, measured in the gaseous state under the same conditions of temperature and pressure, is generally less than the total bulk of the constituent gases measured separately.

In a very few cases the volume occupied by the united substances is the same whether they are combined chemically or only mixed together.

Gay-Lussac's Law of Volumes.

Thus:
- 1 volume of hydrogen with
- 1 volume of chlorine

mixed together would of course form only two volumes of the mixture, but when united they also produce

- 2 volumes of hydrochloric acid gas.

On the other hand,
- 1 volume of oxygen with
- 2 volumes of hydrogen

contract into
- 2 volumes of water vapour.

This will be again referred to in connection with the methods employed for the determination of molecular weights, and until those subjects have been discussed the student must be content to take upon trust all statements concerning molecular volumes, weights, and formulæ.

This law was first enunciated by Gay-Lussac very soon after the laws of reciprocal and multiple proportions by weight had been established by Dalton. We now see the connection between them and the dependence of the one upon the other, though this relation, from want of accurate knowledge, was not at first admitted by Dalton and other chemists. Originally a fact, depending upon experimental evidence only, this statement of Gay-Lussac's now stands in the more important position of a logical deduction from another law, the law of Avogadro (Chap. VII. p. 51).

The results of analysis before Dalton's time were represented by numbers which expressed the proportion of each constituent in 100 parts of the compound. For example, in 100 parts of red oxide of copper there are—

Copper	88·8
Oxygen	11·2
	100·0

In black oxide of copper :—

Copper	79·87
Oxygen	20·13
	100·00

In carbonic oxide :—

Carbon	42·85
Oxygen	57·14
	100·00

In carbonic anhydride :—

Carbon	27·27
Oxygen	72·72
	100·00

Dalton discovered reciprocal and multiple proportions by stripping from them the disguise in which this mode of representing composition enveloped them. For by taking the proportion of some one of the elements in a series of similar compounds as unity, and ascertaining by calculation the proportions which the others bear to it, it is easy to show that these proportions are simple multiples one of another.

In the case of the two oxides of carbon, the composition of which has just been stated, if we calculate the ratio of the oxygen to the carbon in both of them, we find the proportion in the second double that in the first. Thus, taking the carbon as unity—

Carbon.		Oxygen.			Carbon.		Oxygen.
42·85	:	57·14	::		1	:	1·33
and 27·27	:	72·72	::		1	:	2·66

Then plainly

1·33	:	2·66	::	1	:	2

In other words, the proportion of oxygen by weight in carbonic anhydride gas is double that contained in carbonic oxide. Obviously the same fact may be expressed in another way by taking the oxygen as unity instead of the carbon :—

Oxygen.		Carbon.		Oxygen.		Carbon.
57·14	:	42·85	::	1	:	·750
72·72	:	27·27	::	1	:	·375

Then—

| ·750 | : | ·375 | :: | 2 | : | 1 |

Or we may say that carbonic anhydride contains half as much carbon as carbonic oxide. Similarly it will be found that the ratios of the oxygen in the several oxides of nitrogen, of sulphur, and of many of the metals are as 1, 2, 3, &c., to 1, 2, 3, &c., of the other constituent.

Looking about for an explanation of these observations of his, it occurred to Dalton to resuscitate the ancient theory of the limited divisibility of matter. According to this view, matter is made up of definite small parts or particles, called 'atoms,' because it was assumed that they could not be in any way cut or divided. This assumption of finite particles as an explanation of the phenomena of chemical combination thus became the basis of the modern theory of molecules. But Dalton's term 'atom' was applied indiscriminately to the ultimate particles of compound as well as elementary bodies. The relative weights also assigned to these atoms were inferred from the results of the analysis of only a few compounds. Consequently the word 'atom,' in the Daltonian system, may be interpreted sometimes 'molecule,' sometimes 'atom,' and in other cases it corresponds to the term 'equivalent,' as those words are now understood by chemists.

CHAPTER XI.

EQUATIONS.

CLASSIFICATION OF REACTIONS.

Equations.—Chemical changes involve neither the destruction nor the creation of matter, but simply a redistribution of the materials of which the acting masses are composed.

In order, therefore, to represent symbolically the results of any given action it is only necessary to write down the formulæ of the bodies engaged, and then to transpose their symbols in such a manner as to build up the formulæ of the bodies which are produced. We thus arrive at equations in which the signs $+$, $-$, and $=$ are employed, so far as the weights of matter are concerned, in the same sense as in algebra.

The following examples will serve to show the manner in which chemical equations are to be read, as well as the mode of investigating the relative weight of the bodies which are formed or decomposed. The student should practise reading equations aloud, or writing them out at full length, according to these instructions:—

Equation given,

$$2HgO = 2Hg + O_2.$$

This means that two molecules, or 432 parts, by weight, of mercuric oxide yield two molecules, or 400 parts, by weight, of metallic mercury, and one molecule, or 32 parts, by weight, of oxygen.

Young students, however, would do well to ensure precision by filling up such a scheme as the following:—

	$2HgO$	$=$	$2Hg$	$+$	O_2
No. of molecules	2	give	2	and	1
Name	Mercuric oxide		Mercury		Oxygen
Wt. of 1 mol.	$200 + 16 = 216$		200		32
Whole weight	used 432		obtained 400		32

When gases or vaporisable bodies occur in such an equation we can express the volume of each by recollecting that, according to the law of Avogadro, the volumes of gaseous molecules under like conditions are equal. We may also connect together the weight and volume of a gas by applying a rule which will be more fully explained in the next section, namely, that the relative density of a body in the gaseous state [1] is the half of its molecular weight. The weight of a molecule of carbonic anhydride, for example, is 44 : its relative density is, therefore, 22, and the weight of a litre of the gas is 22 times as great as the weight of a litre of hydrogen measured at the same temperature and pressure. Now, the weight of a litre of hydrogen at 0° C., and under a pressure equal to 760 mm. of the barometric column, is ·0896 gram or 1 crith. Hence we may say that a litre of carbonic anhydride weighs 22 criths, or, if the weight is to be expressed in grams, it will be 22 × ·0896 grams.

Example.
$$Cl_2 + H_2 = 2HCl.$$

A molecule of chlorine and a molecule of hydrogen produce two molecules of hydrochloric acid.

	Cl_2 +	H_2	=	$2HCl$
No. of molecules	1	1		2
Name	Chlorine	Hydrogen		Hydrochloric acid or hydrogen chloride
Wt. of 1 molecule	71	2		36·5
Whole weight	used 71	2		obtained 73
Volume	,, 1 or 2	1 or 2		obtained 2 or 4

If, now, we choose to attach a concrete value to the symbols, we express it thus :

$$Cl_2 + H_2 = 2HCl$$
Weight 71 criths   2 criths   73 criths ;

[1] The density of hydrogen being taken as 1.

or

71 × ·0896 grams + 2 × ·0896 grams = 73 × ·0896 grams
Vol. 2 litres + 2 litres = 4 litres.

The student having thoroughly mastered the foregoing examples, and worked some of the exercises given at the end of the section, is now in a position to solve such problems as the following, which may be taken as representative:—

1. How many pounds of zinc are required to make 500 pounds of the crystallised sulphate?

Formula of one molecule of crystallised zinc sulphate, $Zn\,SO_4\,.\,7H_2O$.

Molecular weight 287.

The symbol of zinc, $Zn = 65$, occurs only once in the formula of the sulphate. Hence,

In 287 lbs. of zinc sulphate there are   65 lbs. of zinc.

1 lb. ,, ,, ,, $\dfrac{65}{287}$ lb. of zinc.

∴ 500 lbs. ,, ,, would require $\dfrac{65}{287}$ × 500 lbs.

Ans. $\dfrac{65 \times 500}{287}$ = 113 lbs. 3⅝ oz. of zinc.

2. How many litres of oxygen (at standard temperature and pressure) are obtained by heating 10 grams of potassic chlorate?

$$2KClO_3 = 2KCl + 3O_2,$$

or more simply

$$KClO_3 = KCl + 1\tfrac{1}{2}O_2$$
$$126\cdot6 \quad 74\cdot6 \quad 48.$$

Taking the potassium chlorate in criths, we should obtain 3 litres of oxygen gas.

Then—

$$122\cdot 6 \text{ criths}$$

or $122\cdot 6 \times \cdot 0896$ grams give 3 litres of oxygen.

1 gram would give $\dfrac{3}{122\cdot 6 \times \cdot 0896}$ litres.

∴ 10 grams ,, ,, $\dfrac{30}{122\cdot 6 \times 0\cdot 896}$ litre.[1]

Ans. $\dfrac{30}{122\cdot 6 \times \cdot 0896} = 2\cdot 731$ litres, or 2731 c.c.

CLASSIFICATION OF REACTIONS.

Chemical action may take place in a great many different ways, but every known chemical change to which bodies are subject may be referred to one or other of the following five typical modes of action :—

I. *Combination of entire molecules.*
Examples :

Hg Mercury.	+	Cl_2 Chlorine.	=	$HgCl_2$. Mercuric chloride.
CO Carbonic oxide.	+	Cl_2 Chlorine.	=	$COCl_2$ Carbonyl chloride.
NH_3 Ammonia.	+	HCl Hydrochloric acid.	=	NH_4Cl Ammonium chloride.
CaO Calcium oxide.	+	OH_2 Water.	=	$Ca(HO)_2$ Calcium hydroxide.
PbO_2 Lead peroxide.	+	SO_2 Sulphurous anhydride.	=	$PbSO_4$ Lead sulphate.

---

[1] Taking ·0896 gram as the weight of 1 litre of hydrogen, 1 gram is the weight of 11·16 litres. This number is sometimes useful. Thus, in the example worked out in the text, the calculation would be a trifle shorter, thus :

$KClO_3$ give O_3
122·6 grams   16 × 3 grams
                 or 11·16 × 3 litres.

Then 10 grams of chlorate would give

$$\dfrac{33\cdot 48 \times 10}{122\cdot 6} = 2\cdot 73 \text{ litres.}$$

II. *Splitting up of a compound molecule into its elements, or into simpler molecules.*

Examples:

$CaCO_3$	=	CaO	+	CO_2
Calcium carbonate.		Calcium oxide.		Carbonic anhydride.
H_2SO_4	=	H_2O	+	SO_3
Sulphuric acid.		Water.		Sulphuric anhydride.
$MnCl_4$	=	$MnCl_2$	+	Cl_2
Manganic tetrachloride.		Manganous chloride.		Chlorine.
Hg_2I_2	=	Hg	+	HgI_2
Mercurous iodide.		Mercury.		Mercuric iodide.
$C_7H_6O_5$	=	$C_6H_6O_3$	+	CO_2
Gallic acid.		Pyrogallol.		Carbonic anhydride.

Many familiar decompositions, which at first sight appear to belong to this class, are in reality double decompositions, as an examination of one or two cases will show. The decomposition of mercuric oxide by heat, for example, seems to consist in a simple resolution of the body into its elements,

$$HgO = Hg + O.$$

But when we write the equation molecularly,

$$HgO + HgO = Hg + Hg + O_2,$$

we may see that the decomposition of one molecule of the oxide necessitates the splitting up of another. In order to form one molecule of oxygen—and we believe one molecule to be the smallest quantity of the element capable of independent existence—we must take two atoms of it from two separate molecules of the oxide.

The same remarks apply to other cases, such as the decomposition of potassium chlorate by heat.

III. *Rearrangement of the atoms constituting a molecule so as to give rise to a new body.*

Two examples of this mode of transformation may be given here.

$(NH_4)CNO$ converted by heat into $CO(NH_2)_2$.
Ammonium cyanate.   Urea.

$C_6H_5CH_3HN$ converted by heat into $C_6H_4CH_3H_2N$.
Methyl-aniline.   Toluidine.

The further explanation of this kind of change is postponed to the chapter on 'Isomerism' (Sec. IV.).

IV. *Single Displacement.*—In this kind of change an atom, or group of atoms, contained in a molecule is displaced by another atom or group.

Examples:

$$Zn + 2HCl = ZnCl_2 + H_2$$
Zinc.   Hydrochloric acid.   Zinc chloride.   Hydrogen.

$$MgBr_2 + Cl_2 = MgCl_2 + Br_2$$
Magnesium bromide.   Chlorine.   Magnesium chloride.   Bromine.

$$2C_5H_{11}ONO + SO_2 = (C_5H_{11}O)_2SO_2 + 2NO$$
Amyl nitrite.   Sulphur dioxide.   Amyl sulphate.   Nitric oxide.

Many cases of precisely the same character are not easy to find. The fact is, the great majority of reactions belong to the next class of *double* decompositions. Even some which seem to be single decompositions must in strictness be so considered; thus, the decomposition of hydrochloric acid by sodium must be represented in this manner:

	NaNa	+	HCl	+	HCl		
or	Na_2	+	$2HCl$	=	$2NaCl$	+	$H_2,$
not	Na	+	HCl	=	NaCl	+	H.

V. *Double Decomposition, or Metathesis.*—This is by far the most general mode of reaction. Two or more molecules coming together exchange some of their constituents so as to give rise to the same number, or to a greater number of molecules. We may, for the sake of completeness, classify double decompositions under three divisions:—

1. Those in which one of the reacting bodies is an element; e.g.

$$\underset{\text{Sodium.}}{Na_2} + \underset{\text{Water.}}{2OH_2} = \underset{\text{Sodium hydroxide.}}{2NaOH} + \underset{\text{Hydrogen.}}{H_2}$$

2. Those in which both are compounds; e.g.

$$\underset{\text{Sodium carbonate.}}{Na_2CO_3} + \underset{\text{Calcium chloride.}}{CaCl_2} = \underset{\text{Sodium chloride.}}{2NaCl} + \underset{\text{Calcium carbonate.}}{CaCO_3}$$

3. Those which result in the formation of 'substitution' compounds. This kind of reaction is not essentially different in its nature from cases 1 and 2, but substitution products among carbon compounds constitute a class of bodies so remarkable in their characters as to deserve special notice. Some examples of their formation are given in this place, more particular mention being reserved for a later chapter.

Examples:

$$\underset{\text{Methane.}}{CH_4} + \underset{\text{Chlorine.}}{Cl_2} = \underset{\text{Chloromethane.}}{CH_3Cl} + \underset{\text{Hydrochloric acid.}}{HCl}$$

$$\underset{\text{Benzene.}}{C_6H_6} + \underset{\text{Nitric acid.}}{NO_2HO} = \underset{\text{Nitrobenzene.}}{C_6H_5NO_2} + \underset{\text{Water.}}{H_2O}$$

$$\underset{\text{Benzene.}}{C_6H_6} + \underset{\text{Sulphuric acid.}}{HO.SO_2.HO} = \underset{\text{Benzene-sulphonic acid.}}{C_6H_5.SO_2.OH} + H_2O$$

$$\underset{\text{Phenol.}}{C_6H_6O} + \underset{\text{Nitrous acid.}}{NO.OH} = \underset{\text{Isonitrosophenol.}}{C_6H_4(N.OH)O} + H_2O$$

CHAPTER XII.

CHEMICAL COMPOUNDS DISTINGUISHED FROM MIXTURES.

WHEN two bodies unite together to form a chemical compound, they merge so completely one into the other as to be no longer recognisable by any physical character. The properties of chemical compounds are always quite distinct from those of their constituents, and in general have not even a remote resemblance to them. In the glistening red crystalline powder called by chemists mercuric oxide and vulgarly 'red precipitate,' no trace can be detected by the eye, or by any other sense, of the liquid, volatile, silvery metal mercury, and the colourless, gaseous oxygen into which it is resolved by the action of heat. In water, again, whether examined in the condition of solid ice, liquid water, or vaporous steam, we should look in vain for any resemblance to the gaseous elements, hydrogen and oxygen, of which it consists. Neither can we detect in the properties of water any that can be regarded as intermediate between those of the two constituents, or such as we should expect to find exhibited by a mixture in which each element retained its independence.

The law of definite proportions furnishes another criterion by which the character of a body under examination may be judged as to its title to rank as a definite chemical species. For when it is found that the elements contained in a substance are united together in the ratio of their atomic weights, the probability that they are chemically combined is very great; whilst if they are not present in such proportion, the inference that the substance is a mixture is certain.

In order to decide whether a given substance is a true chemical compound or a mere mechanical mixture, various considerations are employed by chemists, the nature of which depends very much upon the circumstances of each particular case. If a solid body is the subject of investigation, it is examined under a microscope, in order to see if its appearance is uniform throughout ; or, if crystallisable, it is recrystallised, and the crystals compared with those of the original substance. If soluble in any liquid, it may be treated with a quantity of the solvent insufficient to take up the whole. The part dissolved, after getting rid of the solvent by evaporation or otherwise, ought to agree in every respect with the undissolved portion, if the original body is one compound and not a mixture.

In the case of those liquids which are volatile, and which bear the application of heat without decomposition, the boiling-point should remain constantly at the same temperature during the distillation of the whole, and portions taken from the retort and from the distillate in the receiver ought, in the case of definite compounds, to correspond in density and all other physical and chemical characters.

When the body to be examined is a gas, the action of solvents is tried; and if, after such treatment, the relative proportions of the ingredients are undisturbed, the body may be regarded as probably consisting of one compound. This may be confirmed by observing whether these proportions agree with the combining weights of the elements present. The phenomena of gaseous diffusion are not unfrequently useful in helping to decide whether the elements in a given gas are chemically combined or mechanically mixed. Other means of a mechanical nature may be resorted to in special cases, and occasionally considerable ingenuity is called for in devising methods suited to the occasion.

One or two examples will render these matters more intelligible to the student. We will select cases in which the law of definite proportions would afford no assistance in the solution of the problem.

Fifty-six parts of iron would unite with thirty-two parts by weight of sulphur; but the two elements may be mixed together in the state of fine powder, without exerting upon each other any chemical action whatever, the compound, ferrous sulphide, which would be formed by their union in these proportions being produced only when they are strongly heated together. The *mixture* of these two bodies, though it might be indistinguishable from the compound by appealing to the proportions of the two ingredients, would yet be easily recognised by such properties as the following:—Under a microscope particles of iron and particles of sulphur would be visible; a magnet would withdraw the iron from the powder, and leave the sulphur; carbon bisulphide would dissolve the sulphur, but would not affect the iron; a separation could be effected by merely stirring up in water, when the iron, by reason of its greater density, would sink quickly to the bottom, leaving the sulphur suspended; diluted sulphuric acid poured upon the mixture would evolve hydrogen gas. The chemical *compound* has a uniform appearance under the microscope; if reduced to powder, it could not be divided into two different portions by the use of a magnet, by any solvent, or by elutriation with water; and, lastly, the action of diluted sulphuric acid would result in the evolution of hydrogen sulphide gas, easily distinguishable from hydrogen by its odour, by its solubility in water, and by many other properties.

The domain of organic chemistry supplies numerous problems of the kind we are considering.

In certain kinds of tartar there exist the potassium salts of two acids, which have the same composition, but somewhat different properties. These acids are called respectively tartaric and racemic acids. The former rotates the plane of polarisation of a ray of polarised light to the right, whilst the latter is optically inactive. Crystals of tartaric acid are permanent in the air, whilst those of racemic acid contain a molecule of water, which, escaping at ordinary temperatures, renders the crystals efflorescent. Racemic acid and calcium racemate are decidedly, though not very greatly, less soluble than the corresponding tartaric acid and calcium tartrate. Hence it will be perceived that whilst it is perfectly easy to distinguish the pure acids from each other, the mere estimation of the amount of carbon and hydrogen, or even an examination of a great many of the salts, would not suffice to decide between the one or the other of them and a mixture of the two.

A mixture of alcohol, water, and ether might be made in such proportions (46 : 18 : 74) that it would possess exactly the same composition as alcohol. But such a mixture would be at once distinguished from alcohol by its peculiar odour, and by separating into two layers on addition of water. When distilled, it would be found to boil at a much lower temperature than the boiling-point of alcohol; and after about three-fourths had passed over into the receiver, the liquid left behind in the retort would no longer smell of ether, and would be easily recognised as weak alcohol.

Again, the commercial liquid alkaloid toluidine is an oil which, when distilled, boils steadily at about 200°, and no difference of composition can be detected between the first portions of the distilled liquid and the last. And yet this substance is a mixture of two alkaloids of the same composition, one liquid, the other solid, the boiling-points of which (200° and 198°) differ so slightly that they cannot be separated by any kind of fractional distillation. The only process by which they can be completely separated depends upon the fact that the oxalate of solid toluidine is much less soluble in ether than the oxalate of the liquid base. By patient repetition of this treatment with ether two salts are obtained, which, when recrystallised any number of times, undergo no further change in crystalline form and solubility. In this condition they are believed to be pure and homogeneous.

Examples of mixtures of gases, presenting the same composition as true chemical compounds, might be easily multipled.

Thus, equal volumes of hydrogen and chlorine constitute a gaseous mixture which exhibits the colour and bleaching action of chlorine; and after shaking up with solution of soda, just one half its bulk of colourless inflammable hydrogen remains. Hydrochloric acid gas, which contains the same elements combined in the same proportions, is, on the contrary, a colourless gas which no longer possesses the bleaching power of chlorine, and is readily and completely soluble in water or in solution of soda.

Ethane and hydrogen gases in equal volumes furnish a mixture which would be indistinguishable from marsh-gas by ordinary quantitative analysis. But recollecting that ethane is more than four times more soluble in water than hydrogen, whilst the rate of diffusion of hydrogen is nearly four times that of ethane gas, it would not be difficult to distinguish the mixture from the compound. Further assistance might be derived from a study of the action of chemical agents upon the two gases.

The case of atmospheric air is one of so great importance, that its consideration demands some attention in this place.

Neglecting accidental constituents, as well as the water vapour and carbonic dioxide, which are always present, the analysis of air from various localities has led to the conclusion that it consists almost uniformly of 20·9 volumes of oxygen with 79·1 volumes of nitrogen. The question whether these two elements are united together chemically has been decided in the negative, in accordance with such considerations as the following :—

1. The most accurate analyses seem to indicate that the proportion of oxygen to nitrogen in the atmosphere is not absolutely uniform, as would be the case if it were a compound.

2. The quantities of oxygen and nitrogen present do not bear any simple relation to the combining weights of those elements.

3. When oxygen and nitrogen are mixed together they show no signs of chemical action by evolution of heat or contraction of volume ; and such a mixture, when due proportions are employed, resembles atmospheric air in every respect.

4. Water dissolves the constituents of the air in unequal proportions, so that, by reason of the greater solubility of oxygen, the air which may be expelled from common water by boiling contains a larger proportion of that element

than is present in atmospheric air.[1] A chemical compound would dissolve, as a whole, without change of composition.

5. When the rays emitted from a given source of heat are transmitted through different gases, it is found that compounds absorb a much larger amount than elementary gases or mixtures of elementary gases. Thus the amount of radiant heat absorbed by nitrous oxide, a colourless and transparent gas, is more than 350 times as great as the amount absorbed by a column of equal length of oxygen, nitrogen, or atmospheric air. Between the absorbent powers of the last three gases no difference can be detected, and the natural inference, therefore, is that they are similarly constituted.

6. The elements may be separated to a certain extent by the mechanical process of diffusion through a porous plate, called, in this case, *atmolysis*.

It has also been discovered by Graham (see Chap. V., p. 44) that gases have the power of penetrating thin sheets of india-rubber, and that the rate at which oxygen passes through this material is more than two and a half times that of nitrogen. Upon this observation he has based a very instructive experiment, which proves conclusively the fact that the oxygen of atmospheric air is not combined with the nitrogen. An air-tight india-rubber bag is exhausted as completely as possible by the Sprengel air-pump. When the exhaustion is nearly perfect, it is found that gas can still be slowly extracted from the bag by continuing the operation; and this gas is found by analysis to consist of a mixture of nitrogen and oxygen, containing upwards of forty per cent. of the latter ingredient. The gases thus withdrawn from the bag result from the passage of the gases of the atmosphere through the india-rubber partition; the oxygen, however, more rapidly than the nitrogen. The explanation of this dialytic passage of gases through the apparently impermeable caoutchouc appears to be that the

---

[1] See also Chapter II., p. 28, Solubility of Gases.

gases are absorbed by the external film of that material, that they penetrate in this condition to the other side of the sheet, where evaporation occurs in consequence of exposure to an atmosphere of very slight density.

CHAPTER XIII.

NOMENCLATURE.

A NAME may be used either for the purpose of indicating some particular person or object, or it may serve to point out relationships and to define the position which a thing holds in some system of classification.

In the early days of chemistry the number of different bodies known was comparatively small, and mere indicative names fulfilled all the requirements of the time. But when chemical research began to be regularly followed and crowds of new compounds were constantly presenting themselves, it became necessary to devise names which would serve, not merely to distinguish one compound from another, but to indicate, at least in some degree, the relationship subsisting between allied bodies. The first attempts of this kind were naturally imperfect, the devices employed being wholly inadequate to the requirements of the case. For example, it soon became evident that the mere employment of adjectives, as in the names *blue*, *green*, and *white* vitriol, and the like, could have only a very limited application.

It was only after oxygen had been discovered, and its compounds were called *oxides* by Lavoisier, that chemical names began to assume some appearance of precision. The nomenclature adopted by the leading chemists of that period has met with very general acceptance; and, although modified in detail, the system of the present day is based essentially upon the same principle. The names now employed by chemists for scientific purposes sometimes assume rather formidable dimensions, but, unlike the fanciful names in use in connection with some branches of natural history, every syllable has a significance of its own. And, in spite of their length and frequent uncouthness, it may fairly be claimed for these names that they are in few cases inconvenient practically, whilst they do very fairly realise the idea originated by Lavoisier, namely, that of representing, as by a formula, the composition of the bodies for which they stand.

Nevertheless, many of the old names dating their origin from the times of the alchemists have become, by long familiarity, so incorporated into the language of medicine, of commerce, and the arts, that it is neither possible nor advisable for the chemist to reject their use. In the majority of cases, indeed, they are very serviceable, and no chemical student should disdain to make himself acquainted with such terms as *caustic potash, alum, borax,* or *oil of vitriol wood-spirit* or *marsh-gas,* and to use them when they express all that is required, alternatively with the more formal and systematic names, the employment of which, for certain purposes, has been rendered necessary by the advance of knowledge.

Names of Elements.—Several of the metals—iron, copper, lead—were known in very early times. Those elements which have been discovered by more modern chemical research have received names which in some cases recall their origin, *e.g.* silicon (*silex,* flint) ; whilst others were suggested by some prominent characteristic, *e.g.* chlorine (χλωρὸς, green), or by their chemical relations, real or supposed, *e.g.* oxygen (ὀξύς, acid ; γεννάω, I generate).

Metals which have been discovered in modern times have all been designated by Latinised names, with the termination 'ium'[1]—potassium, sodium, lithium, thallium,

---

[1] Unfortunately this termination has been erroneously applied to the name of the non-metal selenium, from a belief entertained at the time of

&c.; whilst the names of the non-metallic elements are characterised by no syllable which is common to them all.

Names of Binary Compounds.—When two elements are combined together, the last syllable or two of the name of one of them is changed into the suffix *ide*. Thus, we have CaO, calcium oxide; HCl, hydrogen chloride; PbS, lead sulphide. In nearly all cases it is the name of the more negative or chlorous element which is thus modified, the name of the metal or corresponding element remaining unaltered.

Prefixes from the Greek—mono-, di-, tri-, &c.—serve to indicate the number of atoms of the chlorous element present in a molecule of the compound. Thus—

N_2O is nitrogen monoxide.
N_2O_2 ,, nitrogen dioxide.
N_2O_3 ,, nitrogen trioxide.
N_2O_4 ,, nitrogen tetroxide.
N_2O_5 ,, nitrogen pentoxide.

Not unfrequently, however, when the actual number of atoms is doubtful, or when it is desired simply to indicate a relation between two compounds, that which contains the larger proportion of the chlorine or oxygen or similar substance is distinguished by the suffix *ic*, whilst the other ends in *ous*. Thus, in the foregoing series, we may distinguish—

N_2O as nitr*ous* oxide.
N_2O_2, or more correctly NO, as nitr*ic* oxide.

Similarly—

N_2O_3 is nitr*ous* anhydride.
N_2O_5 is nitr*ic* anhydride.

A few other marks serve in special cases; thus, Fe_2O_3,

its discovery that it was a metal. I have ventured in these pages to change the final syllable into *on*, selenion. The alteration of the name silicium into silicon will, I hope, be considered sufficient precedent.

Cr_2O_3, Mn_2O_3, Al_2O_3, are often called *sesqui*oxides.[1] When there is a series of oxides, chlorides, or sulphides of the same elements, the syllables *proto*[2] and *per*[3] are sometimes prefixed to the name to indicate respectively the poorest and the richest in oxygen, chlorine, or sulphur, as the case may be. Thus, the two oxides of iron, which are usually represented by the formulæ FeO and Fe_2O_3, may be distinguished by one or other of the following pairs of names, according as it is desired simply to indicate that the ratio of oxygen to iron is greater in the one compound than in the other; or to imply, what is a matter of far less certainty, that so many atoms of the two elements form one molecule of the compound.

FeO, Ferrous oxide : Iron protoxide ; Iron monoxide ;
                                  or   or
                     Protoxide of iron. Monoxide of iron.

Fe_2O_3, Ferric oxide : Iron peroxide ; Iron sesquioxide ;
                                  or   or
                     Peroxide of iron. Sesquioxide of iron.

Names of Acids and Salts.—The same principles guide the construction of the names of these compounds, and a single example will go far towards explaining their application. It happens, in the case of chlorine, that an unbroken succession of compounds is formed by the union of this element with hydrogen and oxygen. These are their names and formulæ :

        HCl . . . Hydrochloric acid.
        HClO . . . Hypochlorous acid.
        $HClO_2$ . . . Chlorous acid.
        $HClO_3$ . . . Chloric acid.
        $HClO_4$ . . . Perchloric acid.

Here, again, the terminations *ous* and *ic* serve to indicate

---

[1] Latin, *sesqui*, one and a half. [2] Greek, πρῶτος, the first.
[3] Greek, ὑπέρ, above, over, exceeding.

different grades of oxidation, whilst the prefixes *hypo* and *per* respectively announce a smaller and a larger amount of oxygen than is contained in the chlorous and chloric acids. In the name hydrochlor*ic* acid, it is evident that the termination has nothing to do with the presence or absence of oxygen, and this use of the termination forms an exception to the general rule. If we consent to regard acids as salts of hydrogen, we may write names for this series of compounds which are constructed in all respects in the same manner as those which are applied to salts in general.

Instead of	We may write
Hydrochloric acid	Hydrogen chloride.
Hypochlorous acid	Hydrogen hypochlorite.
Chlorous acid	Hydrogen chlorite.
Chloric acid	Hydrogen chlorate.
Perchloric acid	Hydrogen perchlorate.

And here another rule must be attended to, namely, that when the name of a salt ends in *ite*, the name of the acid or hydrogen salt with which it corresponds terminates in *ous*, whilst salts in *ate* are derived from acids whose names end in *ic*.

One more example will serve to emphasise this rule:

$\begin{cases} H_2SO_3 & \text{Hydrogen sulphite, or sulphurous acid.} \\ NaHSO_3 & \text{Sodium-hydrogen sulphite,} \\ & \text{or acid sulphite of sodium.} \\ Na_2SO_3 & \text{Sodium sulphite,} \\ & \text{or disodium sulphite (neutral).} \end{cases}$

$\begin{cases} H_2SO_4 & \text{Hydrogen sulphate, or sulphuric acid.} \\ NaHSO_4 & \text{Sodium hydrogen sulphate,} \\ & \text{or acid sulphate of sodium.} \\ Na_2SO_4 & \text{Sodium sulphate,} \\ & \text{or disodium sulphate (neutral).} \end{cases}$

NAMES OF CARBON COMPOUNDS.

Dr. Hofmann, a few years ago, made a proposition with the object of reducing to some degree of order the confused nomenclature of the very numerous compounds of carbon and hydrogen. The number of atoms of carbon is indicated by incorporating the Greek numeral into the name and introducing a vowel into the last syllable, in order to show the proportion of hydrogen. In the following table the words 'methane, ethane, propane,' are based upon the names of the radicles 'methyl, ethyl, propyl,' which have long been in use and are too familiar to be discarded :—

CH_4 Methane.	CH_2 Methene.				
C_2H_6 Ethane.	C_2H_4 Ethene.	C_2H_2 Ethine.			
C_3H_8 Propane.	C_3H_6 Propene.	C_3H_4 Propine.	C_3H_2 Propone.		
C_4H_{10} Tetrane.	C_4H_8 Tetrene.	C_4H_6 Tetrine.	C_4H_4 Tetrone.	C_4H_2 Tetrune.	
C_5H_{12} Pentane.	C_5H_{10} Pentene.	C_5H_8 Pentine.	C_5H_6 Pentone.	C_5H_4 Pentune.	C_5H_2

The succeeding terms would run : hexane, heptane, octane, enneane, decane, &c.

The radicles formed from these bodies by loss of hydrogen are furnished with the termination *yl*; thus CH_3, derived from methane by removing H, is called methyl ; CH is methenyl ; C_2H_5 ethyl, and so on.

Unfortunately, the names which have long been applied to a great many carbon compounds are in the majority of cases fanciful, and, even in allied compounds, have no relation or resemblance to one another. They generally bear some reference either to the original source of the body, or to some more or less prominent characteristic. Thus, formic acid is so called because it was originally obtained from the bodies of ants (*formica*, an ant), acetic acid because it was procured from vinegar (*acetum*). In like manner succinic acid is the acid obtained by distillation of amber (*succinum*), and lactic acid from sour milk (*lac*).

Attempts have been made to introduce some degree of system into this crowd of heterogeneous materials by restricting the use of certain terminations. Thus the hydrocarbons of the aromatic series have names which all end in *ene*; thus, benzene, toluene, anthracene. Alcohols and bodies resembling them claim the terminal syllable *ol*; *e.g.* carbinol, phenol, thymol. Basic nitrogenous bodies are represented by names ending in *ine*; thus, ethylamine, quinine, strychnine.

These conventions do serve their purpose to some extent, though the whole matter is in a condition far from satisfactory.

EXERCISES ON SECTION II.

1. Give the names of the elements represented by the following symbols: Al, Sb, Fe, Mg, Hg, Mn, Ca, C, Cl, I, N, P, K, S, Ag, Na, Br, Cu, F, H, Pb, O, Si, Zn.

2. Write down the symbols and atomic weights of
   Barium, boron, bromine;
   Calcium, carbon, copper;
   Magnesium, manganese, mercury, silver;
   Phosphorus, potassium, lead;
   Sulphur, sodium, silicon;
   Iron, iodine, chlorine, oxygen, nitrogen.

3. Read these symbols and formulæ, thus:—
N_2 represents one molecule of nitrogen, consisting of two atoms;
O, O_2, OH_2, $2OH_2$, HCl, H_2, Cl_2, NH_3, H_3PO_4, H_2SO_4, $FeSO_4$, $2FeSO_4$, $Al_2(SO_4)_3$, $12OH_2$, $12Al_2(SO_4)_3$, CO_2, $3CO_2$.

4. Write down the formulæ and molecular weights of water, ammonia, hydrochloric acid, carbonic anhydride, sulphuric acid, ferrous sulphate, aluminic sulphate, phosphoric acid.

5. Write down the whole weight represented by each of the following expressions: $2HgO$, $100OH_2$, $3FeS$, $3FeS_2$, $2CS_2$, $KC_4H_5O_6$, $K_2C_4H_4O_6$, $5C_7H_9N$, $12CH_4$, $KAl(SO_4)_2 + 12OH_2$, $3[NH_4Cr(SO_4)_2 + 12OH_2]$.

6. Name the following compounds: BaO, CaO, MgO, ZnS, KCl, NaBr, AgF, H_2S, HI, KCN, SSe, BN, H_3P.

7. BaO, BaO_2; Hg_2O, HgO; FeS, FeS_2; MnO, Mn_2O_3, MnO_2; FeO, Fe_2O_3, Fe_3O_4; N_2O, N_2O_2, N_2O_3, N_2O_4, N_2O_5; P_2S_3, P_2S_5; $SnCl_2$, $SnCl_4$; $FeBr_2$, Fe_2Br_6; Cu_2Cl_2, $CuCl_2$; $CrCl_2$, Cr_2Cl_6, CrF_6; $SbBr_3$, $SbBr_5$.

8. KNO_2, KNO_3 (—ate); K_2SO_3, K_2SO_4 (—ate); KCl, KClO, $KClO_2$, $KClO_3$ (—ate), $KClO_4$; KI, KIO_3 (—ate), KIO_4; $NaHSO_3$, Na_2SO_3; Na_2HPO_4, Na_3PO_4, NaH_2PO_4; H_3PO_2, H_3PO_3, H_3PO_4 (—ic); HCl, HClO, HBrO, $HClO_2$, $HClO_3$, HIO_3, $HClO_4$, $HBrO_4$.

9. Write out the following equations according to the scheme on p. 86:—

   (a) $MnO_2 + 4HCl = MnCl_2 + Cl_2 + 2H_2O$.
   (b) $2KI + Cl_2 = 2KCl + I_2$.
   (c) $SO_2 + 2OH_2 + Cl_2 = H_2SO_4 + 2HCl$.
   (d) $NaNO_3 + H_2SO_4 = NaHSO_4 + HNO_3$.
   (e) $2MnO_2 + 2H_2SO_4 = 2MnSO_4 + 2H_2O + O_2$.
   (f) $2K_2Cr_2O_7 + 8H_2SO_4 = 2K_2SO_4 + 2Cr_2(SO_4)_3 + 8H_2O + 3O_2$.

10. Write out the following equations according to the scheme on p. 87:—

    (a) $2OH_2 + 2Cl_2 = 4HCl + O_2$.
    (b) $CO_2 + C \text{(solid)} = 2CO$.
    (c) $2CO + O_2 = 2CO_2$.
    (d) $2NH_3 = N_2 + 3H_2$.
    (e) $2NH_3 + 3Cl_2 = N_2 + 6HCl$.
    (f) $NH_4NO_3 \text{(solid)} = N_2O + 2H_2O$.

11. Write out in symbolic equations—

    (a) Ammonium chloride and calcium hydroxide give ammonia, calcium chloride, and water.

    (b) Ammonium nitrite (heated) yields nitrogen and water.

    (c) Common salt and sulphuric acid yield sodium, hydrogen sulphate, and hydrochloric acid.

    (d) Copper and nitric acid yield copper nitrate, nitric oxide, and water.

    (e) Mercury and sulphuric acid yield mercuric sulphate, sulphurous anhydride, and water.

    (f) Antimonious sulphide and hydrochloric acid yield antimonious chloride and sulphuretted hydrogen.

12. How many grams of oxygen are required to burn 24 grams of carbon and 32 grams of sulphur?

Exercises on Section II. 107

13. How many pounds of zinc are there in 350 pounds of the sulphate $ZnSO_4$?

14. How much sulphur will give 100 kilograms of sulphuric acid, H_2SO_4?

15. How many pounds of black oxide of manganese are required to yield, by the action of hydrochloric acid, 112 pounds of chlorine?

16. How many pounds of chalk containing 96 per cent. of calcium carbonate, $CaCO_3$, will neutralise 250 pounds of sulphuric acid?

(In the following examples the gases are supposed to be at normal temperature and pressure.)

17. Find the weight of 20 litres of oxygen, of 50 litres of chlorine, of 250 litres of ammonia.

18. How many litres of oxygen are required to combine with

    a. 12 criths of carbon? β. 2 grams of sulphur?
                 γ. 10 grams of carbon?

19. How many litres of chlorine are required to decompose 12 litres of hydriodic acid?

$$2HI + Cl_2 = 2HCl + I_2$$

20. How many litres of chlorine are required for the complete decomposition of 10 litres of olefiant gas?

$$C_2H_4 + 2Cl_2 = C_2 + 4HCl$$

21. How many litres of hydrogen are obtained by dissolving 12 grams of magnesium in an acid?

22. What weight of potassium chlorate is required to yield 35,000 cubic centimetres of oxygen?

23. What materials and what quantities would you employ in order to obtain 50 litres of each of the oxides of carbon?

24. How much mercuric cyanide, $Hg(CN)_2$, must be used to furnish 50 c.c. of cyanogen, C_2N_2, assuming that 60 per cent. of the cyanogen is obtained in the gaseous form?

25. Red oxide of copper contains 88·8 parts of copper and 11·2 parts of oxygen by weight; black oxide of copper contains 79·87 of copper and 20·13 of oxygen. If the formula of the black oxide is CuO, how should the red oxide be represented?

26. Water contains 88·8 of oxygen, 11·1 of hydrogen. If its formula is OH_2, find the formula for peroxide of hydrogen, which contains 94·12 O and 5·88 H.

27. Two hydrocarbons have the following composition:—

	I.		II.
Carbon	85·71 } $= C_2H_4$		92·3
Hydrogen	14·29 }		7·7

Find the formula for II.

28. Show that the composition of the oxides of nitrogen, of manganese, and of chromium are in accordance with the law of multiple proportion.

29. Examine the following equations, attach the name to each formula, and classify the reactions according to Chapter XI., pp. 89 to 92, giving reasons in doubtful cases:—

(a) $Na_2 + 2OH_2 = 2ONaH + H_2$
(b) $Fe + H_2SO_4 . H_2O = FeSO_4 . H_2O + H_2$
(c) (electrolysis) $2HCl = H_2 + Cl_2$
(d) $SO_2 + H_2O_2 = H_2SO_4$
(e) (heat) $+ 3MnO_2 = Mn_3O_4 + O_2$
(f) $OKH + HCl = KCl + OH_2$
(g) $O_2H_2 + O_3$ (ozone) $= OH_2 + 2O_2$
(h) $SO_3 + H_2O = H_2SO_4$
(i) $PH_3 + HI = PH_4I$
(j) $N_2O_5 + H_2O = 2HNO_3$
(k) $P_2O_5 + 3H_2O = 2H_3PO_4$
(l) $2KI + Cl_2 = 2KCl + I_2$
(m) $C_2H_4 + Br_2 = C_2H_4Br_2$
    Ethylene.
(n) $NaNO_3 + H_2SO_4 = HNO_3 + NaHSO_4$
(o) (heat) $2KClO_3 = KCl + KClO_4 + O_2$
(p) $MnO_2 + SO_2 = MnSO_4$
(q) $MnO_2 + 2SO_2 = MnS_2O_6$
(r) $C_2H_4O_2 + 3Cl_2 = C_2HCl_3O_2 + 3HCl$
(s) $K_2S + CS_2 = K_2CS_3$
(t) $C_6H_5HO + HNO_3 = C_6H_4NO_2HO + H_2O$
    Phenol.
(u) $C_2H_4O_2 + PCl_5 = C_2H_3OCl + POCl_3 + HCl$
    Acetyl chloride.
(v) $C_{10}H_{16} + HCl = C_{10}H_{17}Cl$
(w) $3HCl + HNO_3 = 2H_2O + NOCl + Cl_2$
(x) $As_2O_3 + 2HNO_3 + 2H_2O = 2H_3AsO_4 + (NO + NO_2)$
(y) (heat) $NaNH_4HPO_4 = NaPO_3 + NH_3 + H_2O$
(z) (heat) $2Na_2HPO_4 = Na_4P_2O_7 + H_2O$

30. How many cubic centimetres of ammonia (measured at 15° and under 740 mm.) would be obtained from $53\frac{1}{2}$ grams of ammonium chloride?

31. How many cubic centimetres of sulphur dioxide (measured at 20° and 740 mm.) can be obtained by the action of copper on 20 grams of sulphuric acid?

SECTION III.

CHAPTER XIV.

ATOMIC WEIGHTS.

THE first step towards the determination of the ratios, commonly called atomic weights, is to ascertain the proportions in which the several elements unite with some one of their number, the combining proportion of which is either taken as the unit, or whose ratio of combination has already been determined. With this object the form of experiment must be suitably chosen, and every precaution taken against error, which, however, can never be wholly avoided, even by the most skilful and experienced operators.

Oxygen was formerly employed as the standard of comparison, and its atomic weight taken as 100. But, since hydrogen is known to enter into combination in smaller proportion than any other element, it is more convenient to adopt it as the standard, and take its atomic weight as unity. The most important relation to establish at the outset is the proportion in which oxygen and hydrogen unite together to form their most stable and characteristic compound, water. This has been attempted by a number of experimenters, but notwithstanding all the care and labour which has been spent upon the investigation, it can hardly be said that the values obtained represent so accurately the combining ratios of these two elements that further experiment is unnecessary. Two chief methods have been adopted. Dumas, in 1842, published the result of his experiments, the first which were conducted with any near approach to accuracy. His plan consisted in passing pure hydrogen through a weighed bulb containing pure oxide of copper heated to redness, collecting the water thus formed and weighing it. Assuming the value 1 for the atomic weight of hydrogen, Dumas obtained for oxygen the number 15·9607. Erdmann and Marchand by the same process obtained the value 15·9733. The second important method consists in determining directly the relative densities of the two gases by weighing them in large globes. By this mode of operating Regnault found oxygen to be 15·9628 times heavier than hydrogen. On reviewing the work that has been done by other experimenters, these values are substantially confirmed, and it appears to be certain that the combining ratio of oxygen is somewhat less than 16, the number commonly used, and is probably very near to 15·9633.

The most trustworthy estimations of the combining weight of carbon were made by the simple method of burning a weighed quantity of pure carbon, in the form of graphite or diamond, in oxygen, and collecting the carbon dioxide formed by means of caustic potash contained in suitable absorption apparatus, carefully weighed. In carbon dioxide 16 parts of oxygen combine with 6 parts of carbon almost exactly; but assuming the value 15·9633 already given for oxygen, the combining weight of carbon becomes 11·9736.

For the estimation of the combining proportions of other elements different methods have to be adopted. It will be sufficient if two or three of these are indicated. Chlorine, potassium, and silver may be taken as being both important and representative.

By decomposing potassium chlorate by heat or by hydrochloric acid, and weighing the resulting potassium chloride, experiments conducted at different times by Berzelius, Penny, Marignac, and Stas led to results the general mean of which gave 60·846 parts of potassium chloride from 100 parts of potassium chlorate. This may be expressed as follows:—

Potassium chloride : oxygen :: 60·846 : 39·154,
or $\quad$ KCl : O_3 :: 74·4217 : 47·8899.

This number, 74·4217, is the combining weight of potassium chloride, and is made up of the combining weights of potassium and chlorine. These must be separately estimated as follows. Starting with pure silver, it may be converted into chloride either by dissolving it in nitric acid and adding hydrochloric acid, or by heating in chlorine gas. Experiments of this kind by a number of chemists, among whom Stas is conspicuous, have shown that 100 parts of silver combine with 32·8418 parts of chlorine. It has also been shown that 100 parts of silver dissolved in nitric acid require for exact precipitation 69·1032 parts of chloride of potassium. Now, these 69·1032 parts of potassium chloride must contain 32·8418 parts of chlorine (which is the amount required to combine with 100 parts of silver) and 36·2614 parts of potassium. Then, since 69·1032 parts of potassium chloride correspond to 100 parts of silver,

$$\text{KCl} : \text{Ag} :: 74·4217 : x,$$

and x is the combining ratio of silver = 107·696.

When the value already adopted for oxygen, namely, 15·9633, is taken, the combining value for silver deduced from all the best experiments is 107·675. By similar calculations the number for potassium is 39·019, and for chlorine 35·370.

Such numbers as these express, with the greatest approach to accuracy at present possible, the proportion in which these elements enter into certain of their most stable and definite compounds, but the question whether they represent the *atomic* weight, that is, the *smallest* proportions in which the several elements enter into combination, can only be answered by appeal to other considerations. The determination of the atomic weight, then, is accomplished by first estimating as exactly as possible the combining proportion of the element concerned, and then finding, by the application of the methods now to be described, a factor by which this value is to be multiplied or divided. In those cases in which it is possible to apply several of the following rules, it is found that the number indicated by the application of one of them is identical with the number indicated by the others. This is important as showing that when the atomic weights of two elements have been determined by two different methods, the probability that they are comparable with each other and with the same standard amounts to practical certainty. Thus, the atomic weight of silver determined by the specific heat method is equally probable with that of carbon derived from the composition and density of its vaporisable compounds, although silver forms no volatile compound, and carbon exhibits anomalies in its specific heat.

The atomic weight, as deduced from other considerations, is represented in some cases by the same number as the density of the element in the gaseous state, the density of hydrogen being taken as the unit. This is true of oxygen, nitrogen, chlorine, bromine, iodine, sulphur, selenion, and probably potassium and sodium. But the densities of the vapours of mercury, cadmium, and zinc, as compared with hydrogen, are respectively half the atomic weights, whilst the vapour-densities of arsenic and phosphorus are represented by numbers which are twice the atomic weights. Hence the vapour-density of an element is no guide to the determination of the atomic weight.

I. According to the system here adopted, and which will be further dwelt upon in the next chapter, the bulk of one part by weight of hydrogen is regarded as the volume of the atom of that element, and is selected as the unit for comparison of other volumes, atomic and molecular. Twice this bulk of hydrogen contains a molecule, and all molecules in the gaseous state occupy the same volume. Now, according to the atomic theory, a molecule cannot contain less than one atom of any element; and, consequently, if we ascertain what is the smallest quantity of an element contained in a molecule of any compound of which it may be a constituent, we shall have determined the atomic weight of the element.

The atomic weight, then, may be said to be the smallest weight of the element ever found in two volumes of the vapour of any of its volatile compounds, the bulk of one part by weight of hydrogen, at the same temperature and pressure, being considered as one volume.

Suppose, for example, it is required to find by this rule the atomic weight of oxygen, we have only to ascertain the vapour densities of a number of compounds containing that element, and the weight of oxygen contained in each. The results are then tabulated in the following manner:—

Volatile Compounds containing Oxygen	Relative Density, that is Weight, of 1 Volume of Gas or Vapour at same Temp. and Pressure	Weight of Two Volumes	Weight of Oxygen contained in Two Volumes
Water	9	18	16
Carbonic oxide . .	14	28	16
Carbonic anhydride .	22	44	32
Sulphurous anhydride .	32	64	32
Sulphuric anhydride .	40	80	48
Nitrous oxide . . .	22	44	16
Nitric oxide . . .	15	30	16
Alcohol	23	46	16
Ether	37	74	16
Acetic acid . . .	30	60	32
&c.	&c.	&c.	&c.

Two volumes of the vapour of any volatile compound, therefore, never contain less than 16 parts of oxygen, and hence 16, or more accurately 15·96, is accepted as its atomic weight.

This is a rule of very general applicability, for although

a great many of the elements, carbon for example, are quite incapable of being volatilised at any manageable temperature, they yield a large number of easily volatile compounds.

There are, however, many metals which are neither vaporisable by themselves nor when in union with other elements. In such cases this rule cannot be applied, and information has to be sought in a different direction.

II. *Law of Dulong and Petit.*—'The specific heats of the solid elements are inversely proportional to their atomic weights.' Whence it follows that the product of the multiplication of the specific heat by the atomic weight is a constant number.

In the following table are given the specific heats of the most important of the elements, together with their atomic weights :—

Name of the Element	Atomic Weight	Sp. Ht. of Equal Weights	Sp. Ht. of Atomic Weights
Aluminium	27	·2143	5·78
Antimony	120	·0508	6·09
Arsenic	75	·0814	6·11
Bismuth	208	·0308	6·40
Boron (crystallised)	11	·2300	2·53
Bromine (solid)	80	·0843	6·74
Cadmium	112	·0567	6·35
Carbon	12	—	—
α. Wood charcoal	—	·2410	2·89
β. Natural graphite	—	·2020	2·42
γ. Diamond	—	·1469	1·76
Cobalt	59	·1067	6·29
Copper	63·3	·0952	6·02
Gold	197	·0324	6·36
Iodine	127	·0541	6·87
Iron	56	·1138	6·37
Lead	207	·0314	6·50
Lithium	7·02	·9408	6·60
Magnesium	24	·2499	6·00
Mercury (solid)	200	·0319	6·38
Nickel	58	·1092	6·33
Phosphorus	31	—	—
α. Common	—	·1895	5·87
β. Red	—	·1698	5·26
Platinum	195	·0324	6·31

Specific Heat and Atomic Weight.

Name of the Element	Atomic Weight	Sp. Ht of Equal Weights	Sp. Ht. of Atomic Weights
Potassium	39·1	·1655	6·47
Silicon	28	—	—
α. Graphitoidal	—	·1810	5·07
β. Crystallised	—	·1650	4·62
γ. Fused	—	·1380	3·86
Silver	108	·0570	6·16
Sodium	23	·2934	6·75
Sulphur (octahedral)	32	·1776	5·68
Tin	118	·0562	6·63
Zinc	65	·0956	6·23

The greater number of the specific heats given in this table were determined by Regnault. A glance down the fourth column will show that, with three exceptions (boron, carbon, and silicon), the amount of heat required to produce the same change of temperature in the different elements is nearly the same in all cases when the quantities operated upon are in the proportion of their atomic weights. That the numbers representing the atomic heats are not found to be exactly identical is due partly to unavoidable errors in the estimation of the specific heats, and partly to the fact that the different elements are not dealt with under conditions which are strictly comparable with one another. Thus, solid mercury and solid bromine, at the temperatures at which the specific heats were determined, are much nearer to their melting-points than are the solids, copper and iron, at the temperatures at which the same operation was performed upon them. Other circumstances, such as the assumption of different allotropic forms by some of the elements, tend to the introduction of further uncertainty.

Experiments made by F. Weber in 1876 prove that the specific heats of the three exceptional non-metallic elements, carbon, silicon, and boron, increase rapidly with the temperature, and become nearly constant at high temperatures. At 600°, or a little above, the number found for diamond was ·4589, and that for graphite ·4670. These numbers multiplied by 12 give 5·5 and 5·6 respectively as the atomic heat of carbon in these two forms. It will be observed that these numbers are not appreciably lower than those assigned to several other elements of small atomic weights. The law of Dulong and Petit may, therefore, be now regarded as fully established and liable to no exception.

The determination of specific heats being always attended by many sources of error, whilst the combining proportion can be fixed with a very considerable degree of accuracy, the application of this rule, like the first, consists essentially in enabling us to decide as to what multiple of the combining proportion is to be taken as the atomic weight. In doing this we are guided by the formula

$$\text{At. Wt.} = \frac{6\cdot 2}{\text{Sp. Ht.}},$$

where 6·2 is the average atomic heat of a solid element.

Suppose, for example, it is found that 29·5 parts of tin are equivalent to 1 part of hydrogen, and we require to find the atomic weight. The specific heat of tin is ·0562, therefore

$$\text{At. Wt.} = \frac{6\cdot 2}{\cdot 0562} = 110\cdot 3.$$

The atomic weight of tin is, however, not taken to be 110·3, but rather such a multiple of 29·5 as comes nearest to that number, and this is found to be 29·5 × 4 or 118.

III. *Law of Isomorphism.*—If a crystal of common potash alum is immersed in a saturated solution of the purple chrome alum, the purple salt is deposited uniformly over the colourless nucleus, so that the crystal increases in bulk though it undergoes no alteration of form. The resulting crystal may be transferred to a solution of ammonia alum or of iron alum, or manganese alum, and during every fresh immersion it receives a deposit of a different salt upon its surface, the crystalline form, that of the regular octahedron, being throughout preserved.

If instead of thus causing successive layers of the various

alums to be superposed one upon the other, solutions of any two of these salts are mixed together, crystals of the same form are deposited containing the elements of both salts.

The alums are double sulphates, all containing the same amount of water of crystallisation, and having a composition which may be represented by the general formula

$$M' M''' (SO_4)_2 + 12OH_2,$$

in which M' may be Cs, Rb, K, Na, Am, Tl or Ag, and M''' may be Fe, Mn, Cr or Al.

If, therefore, any two of these compounds are compared together, as, for example,

<table>
<tr><td>Potash alum</td><td>$KAl(SO_4)_2.12OH_2$</td></tr>
<tr><td>Soda alum</td><td>$NaAl(SO_4)_2.12OH_2,$</td></tr>
</table>

it is obvious that atom for atom they have the same constitution, but the one contains potassium, the other sodium. This exchange of an atom of one element for an atom of another is in this case effected without producing any alteration in the crystalline structure of the resulting salt, and when bodies thus agree in chemical constitution, and in crystalline form, they are said to be *isomorphous*.

From the examination of a great many instances of the same kind, chemists have been led to infer that when two bodies, composed of the same or similar elements, crystallise in forms belonging to the same crystallographic system, they generally contain the same number of atoms united together in a similar manner.

This statement must be considered to include cases in which groups of atoms (compound radicles) take the place and perform the part of single elementary atoms. The compounds of ammonia with acids, for example, are isomorphous with the corresponding salts of potassium, and a constitution is therefore attributed to the ammoniacal salts similar to that of the potassic salts, the symbols NH_4 being the representative of the metal in these compounds. Thus in the following pairs of compounds there is the most complete concordance in chemical characters as well as in crystalline form:

Cubical.
AmCl (Am = NH_4)   KCl.

Four or Six-sided Prisms (Trimetric).
Am_2SO_4 K_2SO_4

Octahedral (Regular).
$AmAl(SO_4)_2\ 12OH_2$ $KAl(SO_4)_2\ 12OH_2$
Am_2PtCl_6 K_2PtCl_6

Although some of the relations between external crystalline form and chemical constitution are still involved in obscurity, the existence of a great number of well-marked cases of isomorphism is a fact which is familiar to every chemist, and occasionally the application of this principle has led to the settlement of questions relating to atomic weights, regarding which there had been previously more or less of uncertainty.

For instance, alumina, the only known oxide of aluminium, is believed to have the same constitution as ferric oxide, because not only do the oxides themselves agree in crystalline form, but they are capable of replacing each other in their compounds without disturbing the crystalline structure of these bodies. Now, since ferric oxide is universally regarded as a sesquioxide, that is, containing in each molecule two atoms of the metal to three atoms of oxygen, alumina is believed to be formed upon the same type, and if the formula Fe_2O_3 be employed to represent ferric oxide, Al_2O_3 must be admitted as the formula for alumina. If these considerations have to be applied to the determination of the atomic weight of the metal, we have only to refer to the analysis of alumina to find that 100 parts contain

53·3 parts of aluminium,
and 46·7 „   oxygen.

And since, according to the formula, we have 3 × 16 or 48

parts of oxygen united with $2x$ parts of metal, we can easily calculate the value of x ($= 27\cdot 4$), which is the atomic weight.

Other instances of a like character would readily present themselves upon inquiry. The following is an interesting example:—

It is well known that the crystalline forms of sodium nitrate and calc-spar are nearly identical, and a crystal of calc-spar immersed in a saturated solution of the nitrate will grow by uniform deposition of that salt all over its surface. Arragonite (another form of calcium carbonate) is also found in prisms of the same form as common potassium nitrate. These facts tend to prove that calcium carbonate probably contains the same number of atoms as the nitrates of potassium and sodium, and that its formula should be

$$CaCO_3$$
if the others are KNO_3
and $NaNO_3$ respectively.

But this formula cannot be used unless we assume that the atomic weight of calcium is 40, a number which agrees with the value deduced from other considerations.

IV. But independently of the existence of vaporisable compounds and of any application of Avogadro's hypothesis, the atomic weight may in some cases be determined by appeal to purely chemical considerations. Take the case of oxygen again. We know by analysis that 100 parts of water contain 88·88 parts of oxygen to 11·11 parts of hydrogen. These proportions might be expressed by the formula HO (if $O = 8$), or by H_2O (if $O = 16$), or by H_3O (if $O = 24$). To decide between these several values we note the results of decomposing water by chemical agents. Potassium acting upon water produces, first, caustic potash, which contains all the oxygen, together with *half* the hydrogen of the water. By the further action of the metal the second half of the hydrogen is expelled, and potassium oxide is formed.

In this decomposition, then, the hydrogen is displaced in two equal portions, and no reaction is known in which the hydrogen of water is divisible into more than two parts. The oxygen, on the other hand, may be displaced by chlorine, but no such division occurs in this case, and no compound is known in which the whole of the hydrogen in water is associated with half the oxygen and a quantity of chlorine equivalent to the other half. There is no compound intermediate between hydrogen chloride and water.

These facts are expressed in the formula H_2O, where the atomic weight of oxygen is assumed to be 16 (approximately), and the atomic weight of hydrogen is 1. The same value for the atomic weight of oxygen must be admitted in the formulæ for ether and for the anhydrides of monobasic acids, like acetic acid. According to views prevailing up to about 1860, alcohol was hydrated oxide of ethyl, EtO.HO,[1] and acetic acid was hydrated oxide of acetyl, AcO.HO [2] (O being = 8). But the conversion of alcohol into ether, and acetic acid into its anhydride, cannot be readily effected by withdrawing the elements of water.

In the operations by which these transformations are effected an intermediate compound is always formed, which represents a second molecule of the alcohol or acid. Thus, ether may be produced from alcohol by first displacing from it an atom of hydrogen by potassium, as in the case of water, producing the compound EtOKO, and this is converted into ether by making it react with the chloride or iodide of ethyl, EtCl or EtI. So that the resulting ether is EtOEtO.

The series of compounds thus produced is best represented as parallel with the series formed from water, adopting in both cases and for similar reasons the assumption that each molecule contains one atom and not two atoms of oxygen.

[1] Et = C_4H_5, if C is 6.
[2] Ac = $C_4H_3O_2$, if C is 6, and O is 8.

Water, HHO. Alcohol, EtHO.
Potassium hydroxide, KHO. Potassium ethoxide, KEtO.
Potassium oxide, K$_2$O. Ethyl oxide, Et$_2$O.

Again, acetic acid can be deprived of the elements of water only with the greatest difficulty, even by the action of so powerful a dehydrating agent as phosphorus pentoxide; but the anhydride, formerly called anhydrous acetic acid, is instantly produced when acetyl chloride is mixed with acetic acid or a dry acetate. So that the constitution of these compounds is similar to that of the ethyl series just described, thus:—

Acetyl chloride, $\overline{\text{Ac}}$Cl.
Acetic acid, $\overline{\text{Ac}}$HO.
Potassium acetate, K$\overline{\text{Ac}}$O.
Acetic anhydride, $\overline{\text{Ac}}_2$O.

In all these cases the value for the atomic weight of oxygen is necessarily 16 (approx.).

The valuation of the atomic weight of carbon may be effected in a somewhat similar manner. Marsh gas is the simplest compound of carbon and hydrogen known. It contains 3 parts by weight of carbon to 1 part of hydrogen, and assuming the value 3 for the symbol C, this composition could be expressed by the formula CH. By the same hypothesis the formula of carbon perchloride is CCl. But when chlorine acts upon marsh gas it forms three intermediate products; hence the hydrogen in marsh gas is divisible into four equal parts, and the hydrogen must be present in the form of four atoms, and the formula becomes C$_4$H$_4$. If there are four atoms of carbon in a molecule of marsh gas, we must believe that these four remain united together to form a group which remains intact throughout all the vast array of compounds derived from marsh gas, and it would be difficult to assign a reason for the three intermediate stages in the chlorination of marsh gas. So that, since there is no evidence of the divisibility of the carbon in the two oxides of carbon, in marsh gas, in chloroform, and other allied substances, one molecule of each of these compounds is represented as containing one atom of carbon, to which the proportional value 12 is given.

It is evident, therefore, that the valuation of atomic weights by chemical methods also involves the determination of the relative molecular weights of a number of compounds. A more direct consideration of this part of the subject is contained in the next chapter.

CHAPTER XV.

MOLECULAR WEIGHTS AND FORMULÆ.

WHEN a compound has been analysed, it is usual in the first instance to represent its composition by the percentages of the several elements of which it is made up. Acetic acid, for example, contains—

          Carbon . . . . 40·0
          Hydrogen . . . . 6·6
          Oxygen . . . . 53·4

in 100 parts. The next step is to endeavour to write a formula which, whilst expressing the same facts more compactly, gives, at the same time, the number of atoms of the constituent elements, and fixes the relative weight of the molecule.

If the atomic weights of the elements were all equal, the formula would be a mere repetition of the percentages; but since they are different, the number of atoms of the several elements contained in equal weights must be inversely as their atomic weights.

The simplest rule for deducing the formula of a compound from its percentage composition is, therefore, to divide the respective quantities by the atomic weights of the elements.

Thus, if we divide the percentages of carbon, hydrogen, and oxygen in acetic acid by the atomic weights of carbon, hydrogen, and oxygen respectively, we arrive at these results.[1]

$$\frac{40}{12} \ldots \ldots \ldots = 3\cdot3$$

$$\frac{6\cdot6}{1} \ldots \ldots \ldots = 6\cdot6$$

$$\frac{53\cdot4}{16} \ldots \ldots \ldots = 3\cdot3$$

So that evidently there are as many atoms of oxygen present as there are of carbon, and there are twice as many hydrogen atoms. The simplest formula, then, that can be written for acetic acid is CH_2O; but whether this is to be taken as representing a molecule of acetic acid or whether the true formula is some multiple of this, such as $C_2H_4O_2$ or $C_3H_6O_3$, remains to be decided by considerations which now require to be examined.

VAPOUR DENSITY AND MOLECULAR WEIGHT.

According to the law of Avogadro, equal volumes of all true gases, irrespective of chemical composition, contain under the same conditions of temperature and pressure the same number of molecules. It follows from this that the weights of equal volumes must be proportional to the weights of the molecules of which the gases are composed.

This is the principle of the only *direct* method for ascertaining the relative weights of these molecules.

Comparing together, for example, equal measures of hydrogen and hydrochloric acid gases, we find their respective weights represented by the numbers 1 and 18·25. But the weight of hydrogen contained in hydrochloric acid is exactly half the weight of the same element contained in an equal measure of hydrogen, and if we assume that there is

---

[1] See Examples and Exercises, p. 155, No. 25.

one atom of hydrogen in a molecule of hydrochloric acid (and by the theory there cannot be less than one atom), we arrive at the conclusion that the molecule of hydrogen consists of at least two atoms. But, further, we have every reason to believe that whilst the molecule of hydrochloric acid contains at least one atom of hydrogen, it does not contain more than one atom. When metals act upon hydrochloric acid the hydrogen is expelled all at once, and not in several portions, as in the case of water and ammonia. So that, assuming the atomic weight of hydrogen as 1 unit of weight, the molecule of hydrochloric acid must weigh 36·5 units. The weight of hydrogen equal in bulk to this, that is to say, one molecule of hydrogen or two unit volumes, must accordingly be 2.[1]

The same conclusions may be established by a slightly different form of the same argument. Equal volumes of hydrogen and chlorine react to form hydrochloric acid equal in volume to the mixture, or twice the volume of the hydrogen or of the chlorine alone. So that whatever is the number of molecules in the hydrogen gas employed, the matter composing them is distributed into twice the number of molecules of hydrochloric acid. And as there is every reason for supposing that the hydrogen chloride molecules are all exactly alike, each of the original hydrogen molecules are broken up into two equal and similar portions.

Again, two volumes of hydrogen mixed with one volume

[1] The student will now perceive why 2 volumes and not 1 volume or 3 or 4 volumes is regarded as the standard volume of molecules. One volume would be inconvenient, because we find that the molecule of hydrogen can be divided into two equal parts, and this would necessitate the use of fractions. Three volumes would be incorrect, because the hydrogen molecule is divisible into two and not into three atoms. Four volumes would also be inadmissible, because the same bulk, that is one molecule of hydrochloric acid, would then be represented as containing two atoms of hydrogen and two atoms of chlorine, H_2Cl_2. Whereas it is a matter of fact that neither the hydrogen nor the chlorine in hydrochloric acid is divisible into separate parts.

of oxygen, at any temperature above the boiling-point of water, make three volumes of the mixed gas. If a spark be now passed into the mixture the gases ignite, and on cooling to the original temperature leave two volumes of steam—that is, a quantity double the bulk of the oxygen used. Here again the matter composing the oxygen molecules is distributed into twice as many molecules of steam, and hence each of the oxygen molecules must have been divisible into two equal and similar parts.

Regarding the diatomic character of hydrogen and oxygen molecules, other evidence, independent of any hypothesis concerning the physical constitution of gases, is supplied by the following facts.

Metallic copper is capable of expelling hydrogen from hydrochloric acid only very slowly, even when boiled with it. But by adding hypophosphorous acid to a warm solution of sulphate of copper, a brown precipitate of cuprous hydride is thrown down, and this compound, in contact with hydrochloric acid, evolves hydrogen freely and forms cuprous chloride. This reaction can only be explained upon the assumption that it is *the attraction of the hydrogen in the copper hydride for the hydrogen in the acid,* superadded to that of the copper for the chlorine, which determines the metathesis:

$$Cu_2\genfrac{}{}{0pt}{}{H}{H} + \genfrac{}{}{0pt}{}{H-Cl}{H-Cl} = Cu_2\genfrac{}{}{0pt}{}{Cl}{Cl} + \genfrac{}{}{0pt}{}{H-H}{H-H}$$

A great many reactions of a similar character are known, chiefly among compounds containing a relatively large proportion of oxygen, part of which escapes in the gaseous form. Thus, when silver oxide is placed in contact with peroxide of hydrogen, the silver is reduced to the metallic state, water is formed, and oxygen gas evolved.

$$Ag_2O + H_2O_2 = Ag_2 + H_2O + O_2$$

In like manner, permanganic and chromic acids are decomposed by peroxide of hydrogen with evolution of oxygen gas; and in these and similar reactions it has been proved experimentally that half the oxygen comes from the peroxide, half from the acid or other body with which it is in contact. The conclusion seems inevitable that these two halves of the oxygen had an attraction for each other which was sufficient to upset the equilibrium of the unstable compounds of which they previously formed a part.

It is also well known that substances in the *nascent* state—that is, at the instant of their liberation from compounds—are capable of acting far more energetically than when they are employed in the bodily form. Hydrogen *gas*, for example, is generally incapable of combining with other bodies; but when materials such as zinc and dilute sulphuric acid, which are capable of yielding hydrogen, are employed, many decompositions and combinations may be brought about which would be otherwise impossible. Solution of sulphur dioxide may in this way be converted into hydrogen sulphide and water by the action of zinc and diluted hydrochloric acid, the *nascent* hydrogen attacking and combining with both the sulphur and the oxygen. An instructive experiment, illustrating the power of both nascent oxygen and hydrogen, consists in electrolysing a solution of hydrochloric acid coloured with indigo. The liquid in the neighbourhood of both poles is bleached; at the negative, because the hydrogen there liberated combines with the indigo and forms a colourless compound; at the positive, because the chlorine, acting on the water, disengages oxygen, which, whilst still nascent, combines with the elements of the indigo, producing a pale yellow substance. Neither oxygen nor hydrogen in the ordinary gaseous form is capable of producing these effects, which are usually supposed to be due to the superior powers of the atoms whilst still in the free state and before they have partly expended their energies by coupling in pairs or otherwise binding together in molecules.

In order, therefore, to express the molecular weights of gases, we must double the numbers representing their relative densities; and this rule is applied to all gaseous and volatile bodies, with the few exceptions referred to in the next chapter.

The application of this principle to those of the elements that are volatile leads to the conclusion that many of them, like hydrogen and oxygen, consist of molecules having a duplex structure, though examples are not wanting of a more complex as well as of a simpler constitution. We come, in fact, to the following classification:—

	Relative Density	Molecular Weight	Molecular Formula
Monatomic Molecules:—			
Mercury	100	200	Hg
Cadmium	56	112	Cd
Zinc	32·5	65	Zn
Iodine (above 1600°)	63·5	127	I
Diatomic Molecules:—			
Hydrogen	1	2	H_2
Oxygen	16	32	O_2
Nitrogen	14	28	N_2
Chlorine	35·5	71	Cl_2
Bromine	80	160	Br_2
Iodine (below 500°)	127	254	I_2
Potassium	39·1	78·2	K_2
Sulphur (at 900°)	32	64	S_2
Triatomic Molecule:—			
Ozone	24	48	O_3
Tetratomic Molecules:—			
Phosphorus	62	124	P_4
Arsenic	150	300	As_4
Hexatomic Molecule:—			
Sulphur (at 500°)	96	192	S_6

It may be noticed in the table just given that several elements are capable of existing in the gaseous state in the form of molecules of various degrees of complexity. Thus whilst vapour of sulphur at temperatures not much above the boiling-point of the liquid consists of hexatomic molecules, at higher temperatures these are broken up into groups each containing only two atoms. Many compounds undergo a corresponding simplification when heated, and generally it may be stated that even in the gaseous state the number of atoms which remain united to form a molecule is less at high than at low temperatures. This will be referred to in greater detail in the chapter on 'Dissociation.'

SYNTHESIS AND ANALYSIS OF COMPOUNDS.

The cases that we have hitherto had under discussion have been chiefly those of the elements. If we have to examine into the constitution of compounds, we have yet several sources of information which will help in the solution of the problem of their molecular weights and formulæ. One very important kind of argument is deduced from a knowledge of the mode or modes in which the compound may be formed, and of the products of its decomposition under the influence of reagents.

For example, oxalate of sodium is formed when carbonic anhydride gas is passed over heated sodium. The change, however, might be represented by either of the following equations :—

$$Na + CO_2 = NaCO_2$$
$$\text{or, } Na_2 + 2CO_2 = Na_2C_2O_4$$

But the doubt, if there were any, would be resolved in favour of the latter alternative, by the observation that when sodium oxalate is heated it yields carbonic oxide gas, leaving a residue of carbonate.

$$Na_2C_2O_4 = \begin{cases} CO \\ Na_2CO_3 \end{cases}$$

The most direct and rational explanation of this is, that a molecule of the salt contains two atoms of carbon, as represented by the formula.

Again, succinic acid may be built up from its elements by a succession of processes, which are briefly represented in the following series of equations ; and from them we learn, even if no other evidence were forthcoming, that succinic acid contains at least four atoms of carbon in a molecule, and hence must have the formula here assigned to it.

Carbon poles ignited by the electric current in hydrogen gas give rise to acetylene.

$$C_2 + H_2 = C_2H_2$$

Acetylene can be made to combine with hydrogen, yielding ethylene, the known density of which gas renders impossible any doubt as to its molecular weight.

$$C_2H_2 + H_2 = C_2H_4$$

Ethylene combines with bromine thus :

$$C_2H_4 + Br_2 = C_2H_4Br_2$$

This dibromide may be converted into a cyanide :

$$C_2H_4Br_2 + 2KCN = C_2H_4C_2N_2 + 2KBr.$$

Lastly, this cyanide, under the influence of boiling alkali, assimilates the elements of water, and yields up its nitrogen in the form of ammonia.

$$\underset{\text{Cyanide of ethylene.}}{C_2H_4 \begin{Bmatrix} CN \\ CN \end{Bmatrix}} + 2KHO + 2H_2O = \underset{\text{Potassium succinate.}}{C_2H_4 \begin{Bmatrix} CO_2K \\ CO_2K \end{Bmatrix}} + 2NH_3$$

From this salt the acid may be procured by the action of sulphuric acid.

$$\underset{\text{Potassium succinate.}}{C_2H_4 \begin{Bmatrix} CO_2K \\ CO_2K \end{Bmatrix}} + \underset{\text{Sulphuric acid.}}{\begin{Bmatrix} H \\ H \end{Bmatrix} SO_4} = \underset{\text{Succinic acid.}}{C_2H_4(CO_2H)_2} + \underset{\text{Potassium sulphate.}}{K_2SO_4}$$

Evidence is also derivable from a study of products of decomposition. Analysis of glucose, for example, gives proportions of carbon, hydrogen, and oxygen, which may be expressed by $(CH_2O)_n$, a formula identical with that of acetic acid, a totally different substance. The problem is to determine the value of n. When glucose is gently oxidised it yields saccharic acid, a well-established dibasic acid, the two potassium salts of which are respectively represented by the formulæ $KC_6H_9O_8$ and $K_2C_6H_8O_8$. Ammonium saccharate when heated splits up according to the following equation :—

$$\underset{\text{Ammonium saccharate.}}{(NH_4)_2C_6H_8O_8} = \underset{\text{Pyrroline.}}{C_4H_5N} + 2CO_2 + NH_3 + 4H_2O.$$

Further oxidation of saccharic acid yields oxalic acid, $H_2C_2O_4$, and oxalic acid in turn yields carbonic acid.

Hence the molecule of saccharic acid contains C_6, and consequently the molecule of glucose from which it is formed very probably contains at least $C_6H_{12}O_6$, inasmuch as products of oxidation almost always contain either the same number of carbon atoms as the compounds from which they were derived, or a smaller number.

This rule, however, though a very useful guide, is not without exceptions. Thus—

Alcohol, C_2H_6O . yields $\begin{cases} \text{aldehyd,} \quad C_2H_4O \\ \text{and} \\ \text{acetic acid,} \; C_2H_4O_2 \end{cases}$

a propyl alcohol, C_3H_8O ,, $\begin{matrix} \text{propionic aldehyd,} \; C_3H_6O \\ \text{or} \\ \text{propionic acid,} \; C_3H_6O_2 \end{matrix}$

β propylic alcohol, C_3H_8O ,, $\begin{cases} \text{acetone,} \; C_3H_6O \\ \text{or acetic acid,} \; C_2H_4O_2 \\ \text{and formic acid,} \; CH_2O_2 \end{cases}$

Toluene, C_7H_8 . ,, benzoic acid, $C_7H_6O_2$

Xylene, C_8H_{10} . . yields $\begin{cases} \text{toluic acid, } C_8H_8O_2 \\ \text{or} \\ \text{phthalic acid, } C_8H_6O_4 \end{cases}$

&c. &c.

But—

Benzene, C_6H_6 . yields benzoic acid, $C_7H_6O_2$
Phenol, C_6H_6O . ,, first phenoquinone, $C_{18}H_{16}O_4$

In such cases the condensation results from several molecules being simultaneously attacked.

The compound last named, phenoquinone, is known to possess the formula ascribed to it because it is otherwise produced by the union of one molecule of quinone, $C_6H_4O_2$, with two molecules of phenol, C_6H_6O, no other product being formed at the same time.

The determination of molecular weight is, however, more conveniently and directly accomplished either by appeal to vapour density, or, if that is not possible, to one or other of the methods following.

SATURATING POWER OF ACIDS AND BASES.

In the examination of acids and bases, and other bodies which are capable of entering into combination readily, or of suffering the replacement of some of their elements by simple reactions which do not involve destruction of the molecule, the process for determining the molecular weight is generally easy. Suppose, for instance, it is required to determine the molecular weight of sulphuric acid, it is only necessary to add to it various quantities of potash or soda to discover that there are two, and two only, distinct and definite sulphates of potassium and sodium, and that even double salts are possible, in which the two metals figure side by side, in place of the hydrogen of the acid. So that sulphuric acid and its salts are representable by such formulæ as the following :—

$$\left.\begin{array}{l}H\\H\end{array}\right\}SO_4 \quad \left.\begin{array}{l}K\\H\end{array}\right\}SO_4 \quad \left.\begin{array}{l}K\\K\end{array}\right\}SO_4$$

$$\left.\begin{array}{l}K\\Na\end{array}\right\}SO_4$$

$$\left.\begin{array}{l}Na\\H\end{array}\right\}SO_4 \quad \left.\begin{array}{l}Na\\Na\end{array}\right\}SO_4$$

Tartaric acid is a case of similar kind. Its molecular formula cannot be less than $C_4H_6O_6$, on account of the existence of the double tartrates, which prove that the acid contains two replaceable basic hydrogen atoms. Thus :

Cream of tartar	$KHC_4H_4O_6$
Rochelle salt	$KNaC_4H_4O_6 + 4H_2O$
Emetic tartar	$K(SbO)C_4H_4O_6 + H_2O$

Basic derivatives of ammonia are dealt with in a similar way. If we assume that the molecule of hydrochloric acid is represented by the symbols HCl ($= 36\cdot5$), the problem is to find what weight of base will enter into combination with $36\cdot5$ parts of hydrochloric acid so as to produce a neutral compound. In the case of ammonia itself the compound formed with hydrochloric acid is represented by the formula NH_3HCl, in which NH_3 stands for 17 parts by weight of ammonia. In some cases it is practically more convenient to prepare the double salts which the hydrochloride of ammonia and all similarly constituted compounds form with platinic chloride. Ammonio-chloride of platinum has the formula $2(NH_3.HCl).PtCl_4$, and that quantity of the basic body which, in this compound, is capable of taking the place of ammonia, NH_3, is generally taken to be its molecular weight.

This principle is applied especially in the case of those natural alkaloids which contain oxygen and are not vaporisable without decomposition. Quinine, from cinchona bark, is an example of the kind. The formula may be either $C_{10}H_{12}NO$ or $C_{20}H_{24}N_2O_2$. The latter is chosen chiefly because quinine forms two classes of salts, namely, neutral compounds, in which $C_{20}H_{24}N_2O_2$ represents the NH_3 in ammonia salts, and acid compounds, containing half the proportion of base or twice the proportion of acid, as in the following formulæ :—

Neutral
$$\begin{cases} C_{20}H_{24}N_2O_2HCl + 2H_2O \\ (C_{20}H_{24}N_2O_2HCl)_2PtCl_4 + 3H_2O \\ (C_{20}H_{24}N_2O_2)_2H_2SO_4 + 8H_2O \end{cases}$$

Acid
$$\begin{cases} C_{20}H_{24}N_2O_2 + 2HCl \\ C_{20}H_{24}N_2O_2.2HCl.PtCl_4 + H_2O \\ C_{20}H_{24}N_2O_2H_2SO_4 + 15H_2O. \end{cases}$$

It appears, therefore, that the number of atoms of nitrogen in the molecule does not determine the saturating power of the base. This is further illustrated by urea, CH_4N_2O, which forms monoacid salts by union with HCl, with HNO_3, or with $\frac{1}{2}H_2SO_4$, &c. ; also by such bases as rosaniline, $C_{20}H_{19}N_3$, which combines either with HCl, forming the well-known magenta dye, or with 3HCl, forming a brown crystalline salt.

SUBSTITUTION COMPOUNDS.

In other cases, especially when the compound under examination is neutral, and incapable of entering into combination with other bodies of known molecular weight, the results of 'substitution' afford information in the direction required. To take an instance, the hydrocarbon benzene is a neutral liquid, neither acid nor basic, which forms no compounds whose constitution throws any light on the question of its molecular weight. It is volatile, and its vapour density would tell all that we require to know, namely, that to represent a molecule of it we must use the formula C_6H_6 ; but even if we had not this evidence to appeal to, the same result would be indicated by the composition of the products which are formed from it under the agency of chlorine. Acted upon in this way, benzene yields first a product in which one atom of chlorine is combined with six atoms of carbon and five atoms of hydrogen. Now, since this can be shown to consist of one homogeneous chemical compound, and not a mixture, the carbon it contains must be derived from one molecule of benzene, the formula of which therefore becomes C_6H_6. This is confirmed by the succeeding products, the entire series running as follows:—

$$C_6H_6$$
$$C_6H_5Cl$$
$$C_6H_4Cl_2$$
$$C_6H_3Cl_3$$
$$C_6H_2Cl_4$$
$$C_6HCl_5$$
$$C_6Cl_6$$

This removal of hydrogen, and its replacement atom for atom by chlorine or by bromine or iodine, is one of the most remarkable facts in chemistry. For notwithstanding the strong electro-chemical opposition of hydrogen and the halogens, they may be usually exchanged for one another in carbon compounds without altering fundamentally the character of the compound. Acetic acid affords a good instance of this. It yields three chloracetic acids, which like itself are crystallisable, volatile, monobasic acids.

Name	Formula	Melting Point	Boiling Point
Acetic acid	$HC_2H_3O_2$	17°	119°
Monochloracetic acid	$HC_2H_2ClO_2$	62°	186°
Dichloracetic acid	$HC_2HCl_2O_2$	0°?	190°
Trichloracetic acid	$HC_2Cl_3O_2$	53°	195°

One other fact often serves as a useful guide in selecting, for carbon compounds especially, a *minimum* formula. Among the vaporisable compounds in which carbon is united with hydrogen alone, or with hydrogen and oxygen, no example is known in which two volumes of the vapour contain an uneven number of atoms of hydrogen.[1] Consequently if the simplest formula deduced from analysis includes such a number it cannot represent the molecular weight, and must be at least doubled. Thus the composition of phthalic acid is expressed exactly by the symbols $C_4H_3O_2$. But according to the present state of knowledge, a molecule of this kind cannot exist ; the molecular formula cannot be less than $C_8H_6O_4$. Whether this correctly represents the molecule can only be settled by other inquiries, the general nature of which has already been indicated.

But, notwithstanding the multiplicity of rules which serve to guide chemists in the selection of formulæ whereby to represent molecules, there still remain a large number of bodies which cannot be dealt with by any method at present known. Hence many of the formulæ commonly accepted and employed in chemical works are at best expressions of mere guesses enjoying various degrees of probability. Many difficulties occur, for example, among metallic compounds. The formula CrO_3 is generally used for chromic anhydride, not on account of any direct evidence in favour of it, but because of the existence of a volatile oxychloride, CrO_2Cl_2, and the analogy of these two compounds with sulphuric anhydride, SO_3, and the corresponding oxychloride, SO_2Cl_2, also on account of the isomorphism of the chromates and sulphates. Again, potassium permanganate is sometimes expressed by the formula $KMnO_4$, which recalls its isomorphism with the perchlorate, $KClO_4$, but the double formula $K_2Mn_2O_8$ has something to recommend it as satisfying the law of even numbers.[1] So also doubts exist as to the correct mode of representing salts like ferrous and stannous chlorides, there being great probability that the usual formulæ, $FeCl_2$ and $SnCl_2$, should be doubled in order to represent the reacting units or molecules of these compounds.

[1] See Chap. XVI.

CHAPTER XVI.

TYPES—VALENCY.

BODIES which are capable of entering into chemical reactions in the same manner, giving rise under similar circumstances to new products having similar properties, are said to belong to the same type. The idea that the chemical constitution of all known bodies is modelled upon a certain limited number of types supplies a means of classifying them according to their modes of transformation. Thus water is capable, in a variety of ways, of exchanging its oxygen and hydrogen for other elements and groups of elements (compound radicals), and the bodies which result from these exchanges retain, more or less perfectly, the chemical deportment of water, and are said to belong to the water type. Caustic potash, for example, is referred to the water type because, like water, it is capable of exchanging its oxygen for sulphur, and its hydrogen for a metal, for an elementary atom like chlorine, or for a group of atoms, such as ethyl or acetyl. As a memorandum of the correspondence in these transformations, the two series of derivatives may be formulated in a similar manner. Thus:

$$\left.\begin{matrix}H\\H\end{matrix}\right\}O \qquad \left.\begin{matrix}K\\H\end{matrix}\right\}O$$

$$\left.\begin{matrix}H\\H\end{matrix}\right\}S \qquad \left.\begin{matrix}K\\H\end{matrix}\right\}S$$

$$\left.\begin{matrix}H\\Cl\end{matrix}\right\}O \qquad \left.\begin{matrix}K\\Cl\end{matrix}\right\}O$$

$$\left.\begin{matrix}H\\C_2H_5\end{matrix}\right\}O \qquad \left.\begin{matrix}K\\C_2H_5\end{matrix}\right\}O$$

$$\left.\begin{matrix}H\\C_2H_3O\end{matrix}\right\}O \qquad \left.\begin{matrix}K\\C_2H_3O\end{matrix}\right\}O$$

&c. &c.

Formulæ of this kind are often spoken of as *rational*

formulæ. But water is not the only type. Hydrogen, hydrochloric acid, ammonia, and marsh-gas and other bodies are often referred to as types of decomposition, each of which is imitated, more or less closely, by a considerable number of elements and compounds. The student will readily perceive that the list of bodies which might, for special purposes, be selected as types may be extended and varied almost indefinitely.

This plan of registering in the formula some of the facts which have been observed as to the possible transformations of a body has led chemists to infer that it is possible to represent symbolically the relative positions of atoms in space. The student will do well to approach such an idea with caution. Rational or descriptive formulæ of various kinds are not only valuable as memoranda of possible modes of formation and decomposition, but are positively necessary to enable us to distinguish from one another bodies having the same ultimate composition, but different properties. Constitutional formulæ, on the other hand, are very often used with the object of representing the supposed arrangement of the constituents within a molecule. But, although a considerable body of evidence has been accumulated, especially in regard to the numerous compounds of carbon, which almost seems to justify the idea, it should not be forgotten that very little is still known regarding the essential nature of molecules and of their constituent atoms.

Sulphuric acid furnishes a very good example of the kind of fact and argument upon which rational formulæ are based. In this case mere analysis tells us only that the compound contains an atom of sulphur, two atoms of hydrogen, and four atoms of oxygen, or SH_2O_4. But on examination of its salts we find that both the atoms of hydrogen are replaceable by metals; and to indicate this basic function of the hydrogen it is the custom to write it at the beginning of the formula, thus:

$$H_2SO_4.$$

When sulphuric acid is distilled with phosphoric chloride, it yields two products having respectively the formulæ

$$SO_3HCl \quad \text{and} \quad SO_2Cl_2.$$

The former of these bodies results from the removal of an atom of oxygen and an atom of hydrogen from the sulphuric acid, whilst an atom of chlorine is taken up.

$$SO_4H_2 - OH + Cl = SO_3HCl$$

And this exchange is repeated when the second derivative is produced.

$$SO_3HCl - OH + Cl = SO_2Cl_2.$$

Either of the new compounds will reproduce sulphuric acid when dissolved in water. It seems, therefore, that sulphuric acid is capable of breaking up into the groups SO_2 and $2HO$. And this is confirmed by the fact of the production of sulphates by the union of sulphur dioxide with peroxides, as in these instances :—

$$\underset{\text{Lead peroxide.}}{PbO_2} + \underset{\text{Sulphur dioxide.}}{SO_2} = \underset{\text{Lead sulphate.}}{PbO_2SO_2}$$

$$\underset{\substack{\text{Hydrogen} \\ \text{peroxide.}}}{(HO)_2} + \underset{\substack{\text{Sulphur} \\ \text{dioxide.}}}{SO_2} = \underset{\substack{\text{Hydrogen sulphate,} \\ \text{or sulphuric acid.}}}{(HO)_2SO_2}$$

Such reactions as these and many others are recalled when we write the formula—

$$\left. \begin{array}{c} HO \\ HO \end{array} \right\} SO_2$$

or—

$$\begin{array}{c} H \\ SO_2 \\ H \end{array} \Big\} O \Big\} O$$

in which it may be regarded as a derivative from two molecules of water, each of which has lost an atom of hydrogen, so that the two residues are united together by the group SO_2.

And there are many cases in which we write such descriptive formulæ with the utmost confidence that they express possible reactions, although such reactions, for various reasons, may never have been observed.

VALENCY.

The formulæ that we now make use of are, to a great extent, based upon certain assumptions regarding those chemical properties of atoms which are referred to under the name 'valency.'

In the formulæ

$$ClH \quad OH_2 \quad NH_3 \quad CH_4$$

we see one atom of chlorine combined with one atom of hydrogen; one atom of oxygen with two of hydrogen; one atom of nitrogen with three of hydrogen; and one atom of carbon with four atoms of hydrogen; and no compound is known in which one atom of either of these elements—chlorine, oxygen, nitrogen, or carbon—is united with a larger quantity of hydrogen than is represented here.

This difference of combining capacity is further illustrated by the fact that when chlorine is made to act upon water, ammonia, or marsh-gas, the hydrogen contained in one molecule of each of these compounds is distributed into so many separate molecules of hydrochloric acid. Thus:

$$\underset{\substack{\text{1 molecule} \\ \text{of water.}}}{OH_2} + Cl_2 = O + \underset{\substack{\text{2 molecules of} \\ \text{hydrochloric acid.}}}{2HCl}$$

Again:

$$\underset{\substack{\text{1 molecule} \\ \text{of ammonia.}}}{NH_3} + 3Cl = N + \underset{\substack{\text{3 molecules of} \\ \text{hydrochloric acid.}}}{3HCl}$$

The following succession of changes indicates the same thing in the case of marsh-gas:—

1. $CH_4 + Cl_2 = CH_3Cl + HCl$
2. $CH_3Cl + Cl_2 = CH_2Cl_2 + HCl$
3. $CH_2Cl_2 + Cl_2 = CHCl_3 + HCl$
4. $CHCl_3 + Cl_2 = CCl_4 + HCl$

Thus the four atoms of hydrogen which, by the carbon

140   *Chemical Philosophy.*

in the marsh-gas, were united together into one molecule, measuring 2 volumes, are separated into four molecules or eight volumes of hydrochloric acid.

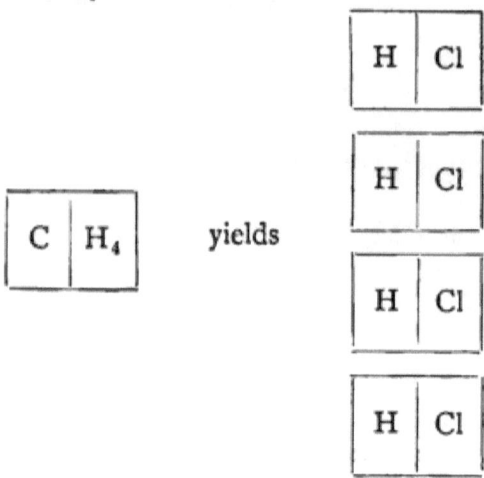

yields

In these cases, therefore, the combining value of an atom of chlorine is equal to that of one atom of hydrogen; it is *univalent.* We summarise these relations when we say that chlorine is a *monad*, because we never find it linked in any gasifiable compound to more than one atom at a time. Hydrogen must also be a monad, for in the production of the compound HCl, whatever is the attraction of the chlorine for the hydrogen, there must be an equal attraction on the part of the hydrogen for the chlorine.

As the result of similar observations, we find that the atom of oxygen is usually diad, of nitrogen triad or pentad, of carbon tetrad, the combining capacity or 'valency' being in each case measured by the number of univalent atoms, such as hydrogen or chlorine, which one atom of these elements can respectively combine with or replace.

The term 'atomicity' is often used to indicate the greatest number of atoms of one kind or another with which a given atom is ever observed to be united. But combining capacity is so much a matter of circumstances that any word which seems to indicate a permanent endowment of the atom, whether as to its form, or motion, or electrical properties, must be objectionable. Many facts seem to point to the conclusion that there is no absolute measure of valency. The capacity of saturation of a given atom depends upon the nature of the elements with which it is associated. An atom of sulphur can take up no more than two atoms of hydrogen, but it is capable of forming a compound with four atoms of chlorine, or with three atoms of oxygen. In like manner, phosphorus forms the compounds PCl_3 and PCl_5, but its affinity for hydrogen extends only to three atoms, PH_3, though a fourth may be taken up if accompanied by an atom of iodine—PH_4I. It has also been observed that the chlorides corresponding with the highest oxides of many of the metals have not yet been produced, and seem to be incapable of existing. Thus there are the oxides CrO_3, UO_3, As_2O_5, Ni_2O_3, but the chlorides Cr_2Cl_6, UCl_5, $AsCl_3$, and $NiCl_2$ indicate the limits of the capacity of these metals for chlorine. It is interesting to notice that in some cases in which the chloride is missing, the corresponding fluoride is known. For example, the fluorides CrF_6 and AsF_5, the representatives of the unknown chlorides $CrCl_6$ and $AsCl_5$, have been described.

Combining capacity or valency is also to a large extent dependent upon temperature, increasing to an unknown maximum with fall of temperature, diminishing with rise of temperature to a minimum which is reached only when the substance becomes a perfect gas. For example, HCl and NH_3 are both gases which, at 500°, have no apparent action upon each other. On cooling they gradually unite to form solid sal-ammoniac. But even then the affinities of one or more of the elements present are not exhausted, for ammonium chloride unites with other chlorides, forming such compounds as the following :—

$$2NH_4Cl.ZnCl_2$$
$$2NH_4Cl.SnCl_4$$
$$2NH_4Cl.PtCl_4 \text{ \&c.}$$

These compounds and sal-ammoniac itself can only exist in the solid state, or perhaps dissolved in water. When heated they yield ammonia and hydrochloric acid gases and other products. The formation of many compounds of this kind leads to the suspicion that the valency of chlorine probably forms the link which binds the other atoms together, and that in the solid state its valency is greater than in its gaseous compounds. The other halogens behave in a similar manner.

Oxygen, again, is apparently diad or bivalent in water and nearly all other vaporisable compounds containing this element. But in many solid compounds, especially in salts containing water of crystallisation, the facts can best be explained on the hypothesis of a greater combining capacity, probably amounting to four units or more.

The variation of valency under varying conditions is analogous to the variation in the number and disposition of the faces of crystals. Every crystallisable substance exhibits a primary or simple form, made up usually of a minimum number of faces and solid angles. Thus common salt habitually crystallises from water in cubes (6 faces), but from a solution containing urea it yields octahedrons (8 faces). Alum usually forms simple regular octahedrons, but occasionally produces dodecahedrons (12 faces), or forms intermediate between cube or octahedron and dodecahedron. Again, many substances, called on this account *dimorphous*, are capable of assuming two distinct crystalline forms belonging to different crystallographic systems, these different forms being produced most commonly by crystallisation at different temperatures. For example, nitre usually crystallises in six-sided prisms of the trimetric system, but when deposited from a hot solution it often forms rhombohedrons of the hexagonal system. In all cases of this kind the one form is more stable than the other, and there is a constant tendency for the less stable to pass into the more stable form.

Combining power, then, is known to be variable. It is,

however, useful to remember the capacity of combination habitually shown by the elements in their more familiar compounds, and those which have been sufficiently studied admit of classification into the six divisions displayed in the following table :—

VALENCY OF THE PRINCIPAL ELEMENTS.

NON-METALS.					
Monads.	Diads.	Triads.	Tetrads.	Pentads.	Hexads.
F	O	B	C	N	S
Cl	S	N	Si	P	Se
Br	Se	P			
I	Te				Te

METALS AND METALLOIDS.					
Ag	Hg	Al	Ir Ro	V	Os Ru
	Cu	Au	Pt Pd	As	W
	Co Mn	In			Mo
	Ni Fe			Sb	
H	Cd	Tl	Cr		
Li	Zn		Mn	B	U
Na	Mg		Fe		
K	Ca	V	Pb		
Rb	Sr	As	Sn	Ta	Cr
Cs	Ba	Sb	Ti	Nb	Mn
		Bi	Zr		Fe
	Pb		Co		
			Ni		

An arrangement of this kind necessarily involves the

separation of many elements which, in properties, are closely allied together, and the association of others which have very little in common. Thus we have to look for thallium among the triads, although it has strong points of resemblance on the one hand to the alkali metals, and on the other hand to lead. Boron, again, is separated from carbon, aluminium from chromium and iron, lead from silver.

CONSTITUTIONAL AND GRAPHIC FORMULÆ.

The compound resulting from the union of two or more atoms is called a saturated compound, when the valency of each atom present is satisfied. This condition is fulfilled when one atom of a monad is combined with another monad, or when two atoms of a monad combine with one atom of a diad, or three of a monad with one of a triad, four monad or two diad atoms with one tetrad, and so forth.

Examples of various orders of compounds are shown in the following formulæ, in which the combining power of each atom present is supposed to be neutralised by that of other atoms. In the notation here introduced, it must be understood that the symbol placed on the left of a formula represents an atom to which all on the same line are directly united. Those also which are connected by a bracket are united together. Thus,

$$\left\{ \begin{array}{l} CH_3 \\ CH_3 \end{array} \right.$$

means that there are two atoms of tetrad carbon united by one-fourth of their combining power, whilst each retains three atoms of hydrogen. The same relations are expressed in this figure, or 'graphic formula:'

$$\begin{array}{c} H \ \ H \\ | \ \ \ | \\ H-C-C-H \\ | \ \ \ | \\ H \ \ H \end{array}$$

Constitutional Formulæ.

Of course the lines connecting the symbols are not designed to represent any substantive bond or link, but merely indicate the manner in which the combining capacity of each atom is disposed of.

Hydrochloric acid.
HCl or $H-Cl$

Water.
OH_2 or $H-O-H$

Copper sulphide.
CuS or $Cu=S$

Hydrogen peroxide
$\begin{cases} OH \\ OH \end{cases}$ or $H-O-O-H$

Phosphoric chloride.
PCl_5 or
$$\begin{array}{c} Cl \\ | \\ Cl-P-Cl \\ / \backslash \\ Cl \quad Cl \end{array}$$

Aluminic chloride
$AlCl_3$ or
$$\begin{array}{c} Cl-Al-Cl \\ | \\ Cl \end{array}$$

Stannic chloride.
$SnCl_4$ or
$$\begin{array}{c} Cl \\ | \\ Cl-Sn-Cl \\ | \\ Cl \end{array}$$

Phosphoric oxide.
$\begin{cases} PO_2 \\ O \\ PO_2 \end{cases}$ or
$$\begin{array}{ccc} O & & O \\ \| & & \| \\ P-O-P \\ \| & & \| \\ O & & O \end{array}$$

Acetic acid.
$\begin{cases} CH_3 \\ CO(OH) \end{cases}$
$$\begin{array}{c} H \quad O \\ | \quad \| \\ H-C-C-O-H \\ | \\ H \end{array}$$

Sulphuric acid.
$SO_2(HO)_2$ or

$$H-O-S(=O)(=O)-O-H$$

Potassium dichromate.
$$\begin{cases} CrO_2(OK) \\ O \\ CrO_2(OK) \end{cases} \text{ or } K-O-\underset{\underset{O}{\|}}{\overset{\overset{O}{\|}}{Cr}}-O-\underset{\underset{O}{\|}}{\overset{\overset{O}{\|}}{Cr}}-O-K$$

LAW OF EVEN NUMBERS.

When several compounds are formed by the union of two elements in different proportions, it is very commonly noticed that the change of valency or combining capacity of the central atom to which the rest may be supposed to be attached takes place by pairs of units.

Phosphorus, for example, forms two chlorides; nitrogen combines with three atoms of hydrogen in ammonia, and with four atoms of hydrogen and an atom of chlorine in chloride of ammonium. Sulphur also yields compounds, in which one atom of that element is combined with two atoms of hydrogen, with two atoms of oxygen, and with two atoms of oxygen and two atoms of chlorine. The formulæ of these compounds are represented as follows:—

$$\begin{array}{lll} PCl_3 & NH_3 & SH_2 \\ PCl_5 & NH_4Cl & SO_2 \\ & & S\begin{cases} O_2 \\ Cl_2 \end{cases} \end{array}$$

It will be observed that the difference in the first case amounts to two atoms of chlorine, which represent two atoms of hydrogen, the unit of valency. In the second case, one atom of hydrogen and one atom of chlorine have been added. In the third, the oxide SO_2 may be taken to represent a hypothetical hydride SH_4, whilst the oxychloride corresponds with the unknown compound SH_6. In each series, the advance in combining power is equivalent to the assumption of two atoms of hydrogen. And so it is in a large number of other cases.

It seems, therefore, that, as a general rule, the index of valency of any given atom is either an even or an odd number; or, as it has been expressed, elements are uniformly either 'artiad' or 'perissad.'[1] And consequently the sum of the indices of valency of all the atoms present is an even number. (See also p. 135.)

But there are not wanting exceptions to these statements; and although the number is at present not great, their marked character is sufficient to destroy much of the apparent significance of those more numerous instances which conform to the general rule.

The following are some of the most notable exceptions:—

Nitric Oxide.—This most remarkable compound is a colourless gas, difficult to liquefy, almost insoluble in water, and unchanged by heat. It exhibits all the characteristics of an unsaturated compound. Thus it unites with oxygen, with chlorine, with sulphuric anhydride, and with many metallic salts.

It is composed of 14 parts of nitrogen with 16 parts of oxygen; its relative density is 15 ($H=1$), and consequently its molecular weight is 30. It therefore contains one atom of nitrogen (perissad), combined with one atom of oxygen (artiad), and thus it breaks the law of even numbers. This difficulty might be avoided by employing the double formula,

$$N_2O_2 \quad \begin{array}{c} N=O \\ | \\ N=O \end{array} \quad \text{or} \quad \begin{array}{c} N-O \\ \| \quad | \\ N-O \end{array} \quad \text{or} \quad \begin{array}{c} N=O \\ ||| \\ N=O \end{array} \quad \text{or} \quad \begin{array}{c} N-O \\ |||| \quad | \\ N-O \end{array}$$

but that its low density and difficult condensability point conclusively to the simpler expression NO as the symbol of its molecule.

Nitric Peroxide.—This compound is described in Chapter

---

[1] ἄρτιος, even, and περισσός, odd.

XVIII. At 150°–200° the formula NO_2 represents two volumes of the gas; but even at much lower temperatures this compound exists mixed with the tetroxide, N_2O_4.

Tungsten and Molybdenum Pentachlorides, WCl_5 and $MoCl_5$.—These compounds afford very remarkable instances of the association of an artiad atom with an uneven number of perissad atoms, and consequent infraction of the law under discussion. Hexchlorides of both these elements corresponding with their trioxides, exist; but these chlorides when heated are split up into pentachlorides and free chlorine. The pentachlorides are volatile without decomposition, and the vapours exhibit normal densities.

COMPOUND RADICLES.

Any portion of a molecule which is capable of being detached and transferred to some other molecule is called a 'radicle,' whether it consist of a single atom or of a group of atoms. The term 'compound radicle' is, however, not usually applied to a group unless it makes its appearance in several different bodies. Compound radicles present different degrees of valency, just as do the atoms of which they are built up, so that they are capable of linking together various proportions of other elementary or compound radicles. This fact may be experimentally verified by such reactions as the following:—

Ordinary disodic phosphate is alkaline to test-paper; silver nitrate is neutral. When these two salts are mixed together in equivalent proportions a yellow precipitate of phosphate of silver is thrown down, whilst the liquid becomes strongly acid. The reason of this is apparent when we express the metathesis in the form of an equation.

Compound Radicles.

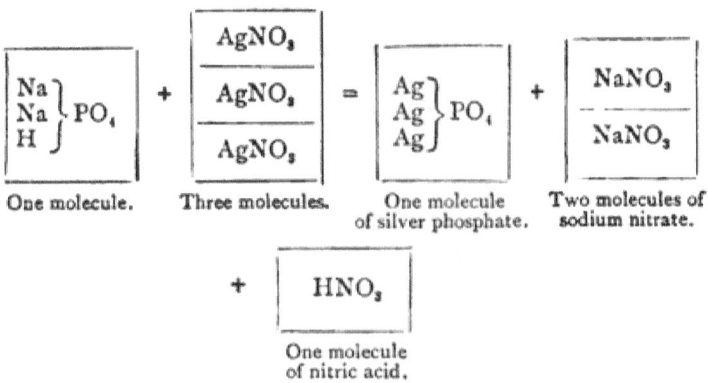

The group PO_4 is trivalent, and so it holds together the two atoms of sodium and one atom of hydrogen in one molecule. But when the interchange occurs, these become respectively united with three (NO_3) groups, each of which is univalent, and incapable of connecting itself with more than one atom at a time. Three new molecules result, one of which is nitric acid, the presence of which can be recognised by test-papers.

Although nearly all unsaturated compounds are capable of entering into combination, and many of them perform the part of well-defined radicles, it does not follow that all radicles should be capable of isolation, and the definition of a radicle by no means involves this idea. On the contrary, we have already examined phenomena (pp. 123 to 126) which indicate that when atoms such as H, Cl, O, or N are liberated from combination, they combine in pairs, and thus satisfy each other's attractions, unless they find themselves in the presence of other radicles with which they can immediately unite. Compound radicles resemble elementary atoms in this respect. None of the following groups, for example, are known in the free state, the formulæ [1] representing semimolecules, or what may be termed chemical atoms of these radicles.

[1] The dashes serve to indicate the usual valency of each group.

Hydroxyl (OH)' contained in acids, alcohols, and metallic hydroxides.
Potassoxyl (OK)' contained in potassium oxysalts.
Cyanogen (CN)' contained in cyanides.
Ammonium (NH$_4$)' contained in the salts of ammonia.
Arsendimethyl $\left(\text{As}^{CH_3}_{CH_3}\right)'$ in kakodyl and its compounds.
Methyl (CH$_3$)' in methylic alcohol and derivatives.
Amidogen (NH$_2$)' in primary amines and amides.
Methenyl (CH)''' in chloroform and similar bodies.

When displaced from any of their compounds they do not remain isolated, but unite in pairs, producing molecules which in some cases are stable enough to maintain an independent existence.

We have, for example—

Radicle Name.	Corporate Name.	
Hydroxyl	Hydric Peroxide	(OH)$_2$ or O$_2$H$_2$
Potassoxyl	Potassic Dioxide	(OK)$_2$ or O$_2$K$_2$
Cyanogen	Cyanogen Gas	(CN)$_2$ or C$_2$N$_2$
Arsendimethyl	Kakodyl	$\left(\text{As}^{CH_3}_{CH_3}\right)_2$ or As$_2$(CH$_3$)$_4$
Methyl	Ethane	(CH$_3$)$_2$ or C$_2$H$_6$
Amidogen	Diamidogen or Hydrazine	(NH$_2$)$_2$ or N$_2$H$_4$
Methenyl	Acetylene	(CH)$_2$ or C$_2$H$_2$

The only free monad radicles known are the two bodies already described, namely, nitrosyl, or nitric oxide, NO, and nitryl, or nitric peroxide, NO$_2$. Instances of free diad radicles are, however, more numerous. Thus we have

Mercury	Hg''
Cadmium	Cd''
Carbonic oxide (carbonyl)	(CO)''
Sulphur dioxide (thionyl)	(SO$_2$)''
Ethylene	(C$_2$H$_4$)''
Ammonia	(NH$_3$)'

Molecules of this kind are of the same order as those referred to at the beginning of the chapter; but why in so many cases the number of unemployed units of valency should be an even number, has not yet been satisfactorily explained.

Accepting the definition of a 'radicle' given on p. 148, it is obvious that there are a great many commonly recognised radicles which can hardly be expected ever to assume a bodily existence apart from the compounds in which they occur associated with elements of a different chemical character.

In all the carbonates, for example, a group consisting of one atom of carbon and three atoms of oxygen occurs, and this group is capable of being exchanged for Cl_2 or $(OH)_2$ or O by double decomposition. It is, therefore, entitled to be spoken of as a compound radicle, although, by reason of the large proportion of oxygen it contains, its condition would be that of unstable equilibrium, even if it could assume temporarily an isolated existence.

Similar remarks apply to such radicles as NO_3 (of nitrates), ClO_3 (of chlorates), SO_4 (of sulphates), PO_4 (of phosphates), and the rest, which, under possible experimental conditions, have never yet been isolated.

But, after all, it is necessary to remind the student that our system of notation is to be understood and employed only in a unitary sense. Every molecule must be regarded as one entire and undivided unit, whose actions and reactions proceed from the resultant of all the different forces exerted by its several constituent parts. A chemical compound may be compared to a musical chord, constituted, doubtless, of many and complex elements, but communicating to the ear the impression of singleness and harmony. And though there are many cases in which one constituent may have a predominating influence in virtue of its mass or special store of unexhausted energy, it never acts as though it were alone, its behaviour being always modified by the presence of the other elements with which it is associated.

The doctrine of radicles, no less than that of valency, and the graphic notation founded upon it, is at present to be regarded solely in the light of a convenient but not absolutely necessary system of recording and comparing facts concerning the changes of composition to which bodies are subject under the influence of chemical attraction.

Note.—The following memoranda will serve to assist the student in writing the formulæ of many common salts. In order to construct any required formula it is only necessary to place a symbol or group of symbols, taken from under the positive sign, side by side with a symbol or group taken from under the negative sign, and to adjust the quantity of each so as to comply with their respective habits of combination.

Thus, let

R' represent a univalent radicle,
R'' ,, ,, bivalent radicle,
R''' ,, ,, trivalent radicle, &c.

Then it is only necessary to remember that

R' combines with R',
$2 R'$,, ,, R'',
or R'' ,, ,, R'',
$3 R'$,, ,, R''',
$3 R''$,, ,, $2 R'''$, &c.

In this way the student will readily learn to compose the unitary formulæ of all the most commonly occurring compounds, without risk of falling into any serious error.

VALENCY OF COMMON SALT RADICLES.

Univalent.		Bivalent.		Trivalent.		Quadrivalent.		Sexvalent.
−	+	−	+	−	+	−	+	+
F	H	O	Ba	PO_4(ortho)	As	SiO_4 (ortho)	Sn (ic)	Al_2
Cl	K	S	Sr	AsO_3	Sb	P_2O_7 (pyro)	Pt	Cr_2 (ic)
Br	Na	SO_3	Ca	AsO_4	Bi			Fe_2 (ic)
I	NH_4	SO_4	Mg	BO_3	Au			Mn_2 (ic)
CN	Ag	S_2O_3	Zn					
HO		CO_3	Cd					
NO_2		C_2O_4	Hg (ic)					
NO_3		$C_2H_3O_6$	Cu (ic)					
ClO		SiO_3(meta)	Pb					
ClO_2		CrO_4	Fe (ous)					
ClO_3		MnO_4	Mn (ous)					
ClO_4		Mn_2O_7	Sn_2 (ous)					
PO_3 (meta)			Hg_2 (ous)					
$C_2H_3O_2$			Cu_2 (ous)					

EXERCISES ON SECTION III.

1. One atom of antimony is said to be equivalent to three, and one atom of zinc to two, atoms of sodium. Explain this statement.

2. Distinguish between atomic, equivalent, and molecular weights.

Give the atomic and equivalent weights of mercury, zinc, chlorine, iodine, sulphur, iron, and copper.

Also write down the molecular weights of H_2S, PCl_5, AsH_3, H_2SO_4.

3. What weight of sulphuric acid can be precipitated by one gram of barium chloride?

4. What is the weight and volume (at normal temperature and pressure) of the hydrogen contained in 10 grams of microcosmic salt, $NaNH_4HPO_4.4H_2O$?

5. Enumerate very briefly the various methods by which atomic weights may be determined; and indicate in the case of each of the following elements the method or methods which would be applicable: oxygen, chlorine, potassium, mercury, carbon, sulphur, lead, silver, arsenic, silicon, barium, copper, manganese.

6. The specific heat of iron is ·1138. State approximately its atomic weight.

7. The specific heat of cadmium is ·0567, and its equivalent 56. Give its atomic weight.

8. The equivalent of platinum is 49·4, and its perchloride has the formula $PtCl_4$. Find its specific heat.

9. The formula of water was formerly written thus, HO, and subsequently, for some years, H_2O_2 (assuming $O = 8$). Discuss both these formulæ, pointing out any inconsistencies you may detect in them.

10. Complete the equation

$$Ag_2O + H_2O_2 =$$

Quote analogous reactions, and explain the theoretical significance of these facts.

11. The volume of the molecule of a compound body in the gaseous state is double the volume of the atom of hydrogen. Examine the truth of this statement; give the experimental facts upon which it is based, and discuss any exceptions to it with which you are acquainted.

12. Acetic acid contains C 40, H 6·6, and O 53·4 per cent.; and chloracetic acid contains 37·5 per cent. of chlorine. Calculate the molecular weight of acetic acid.

13. Explain the signification of the several formulæ for potassic sulphate,

$$K_2O.SO_3 ; \left.\begin{array}{l}KO\\KO\end{array}\right\}SO_2 ; K_2SO_4$$

and

$$\begin{array}{c}O\quad\quad O-K\\ \diagdown\!\!\diagup\\ S\\ \diagup\!\!\diagdown\\ O\quad\quad O-K\end{array}$$

14. What is understood by the terms *valency* and *atomicity* respectively, and how would you ascertain the valency and atomicity of a given element—for example, of carbon or phosphorus?

15. What is the atomicity of each of the following radicles:—

S, O, Cl, OH, NH_4, NH_3, NH_2, NH, N, N_2 PO, SO_2?

16. With the help of the table on p. 152 write down the formulæ of the following salts: sodium fluoride, silver sulphate, mercuric cyanide, mercurous phosphate, barium chlorate, bismuth chloride, ferrous orthosilicate, cupric acetate, ferric nitrate, chromic oxalate, stannic phosphate, calcium hypochlorite, &c. &c.

17. Write the formulæ of nitric oxide, carbonic oxide, white arsenic, acetic acid, benzoic acid, oxalic acid, aluminium chloride, ferric chloride, orthophosphoric acid, and describe briefly the method or methods by which the molecular weight of each has been determined.

18. Taking hydrochloric acid HCl, water $\left.\begin{array}{l}H\\H\end{array}\right\}O$, ammonia $N\left\{\begin{array}{l}H\\H\\H\end{array}\right.$ and carbon dioxide $C\left\{\begin{array}{l}O\\O\end{array}\right.$ as typical, write out the formulæ of about a dozen substances which conform to these types.

19. If the specific heat of a metal is ·03, state approximately its atomic weight. Calculate the exact value of the atomic weight, the composition of the chloride being, metal 2·915 parts to 1 part of chlorine by weight.

20. The atomic weight of silver being 108, and its specific heat ·057, another metal M, of which 70 parts unite with 35·5 parts of chlorine, is found to have the specific heat ·0306. What is the atomic weight of this metal and the formula of its chloride?

21. Give reasons for representing hydrobromic acid by a formula similar to that of hydrochloric acid; HCl, HBr.

Exercises on Section III. 155

22. Alcohol, ether, and acetic ether have the following rational formulæ:

$$\left.\begin{array}{c}C_2H_5\\H\end{array}\right\}O \;,\quad \left.\begin{array}{c}C_2H_5\\C_2H_5\end{array}\right\}O \;,\quad \left.\begin{array}{c}C_2H_5\\C_2H_3O\end{array}\right\}O$$

What arguments could you draw from the existence of such bodies in favour of the number 16 as the atomic weight of oxygen? Why is it probably neither 8 nor 32?

23. The formula of the molecule or chemical unit of ammonia is NH_3. What is the meaning of this formula, and what are the reasons for using it?

24. A compound is found by analysis to have the following composition:—

Carbon	52·18
Hydrogen	13·04
Oxygen	34·78
	100·00

To find its simplest formula.

25. The analysis of a compound leads to these numbers:—

Carbon	37·20
Hydrogen	7·90
Chlorine	54·95
	100·05

It is not often that the formula can be calculated so easily as in the example given in the text and in the last exercise. It must be borne in mind that in actual practice a slight loss is incurred in the estimation of many elements. The number for hydrogen, however, generally comes out a trifle too high. Oxygen is always estimated by taking the difference between the total weight of the body analysed and the sum of the weights of the constituents which have been actually weighed.

In the present example we proceed in the following manner:—
Divide the percentages by the atomic weights in the usual way:

$$\frac{37\cdot2}{12} = 3\cdot1 \text{ atoms of carbon.}$$

$$\frac{7\cdot9}{1} = 7\cdot9 \text{ ,, ,, hydrogen.}$$

$$\frac{54\cdot95}{35\cdot5} = 1\cdot54 \text{ atom of chlorine.}$$

Divide the three quotients by the last, which is the least.

$$\frac{3\cdot 1}{1\cdot 54} = 2\cdot 01 \text{ atoms of carbon.}$$

$$\frac{7\cdot 9}{1\cdot 54} = 5\cdot 12 \text{ ,, ,, hydrogen.}$$

$$\frac{1\cdot 54}{1\cdot 54} = 1 \text{ atom of chlorine.}$$

Now, recollecting that the percentages found by analysis are not exactly true, but only close approximations to the correct numbers, and remembering that the hydrogen is generally in slight excess, we may safely reject the two small fractions which occur in the above numbers, and the formula then reads

$$C_2H_5Cl.$$

To prove that this represents correctly the composition of the body, it is well to recalculate the percentages on the basis of this formula. This calculation is performed in the following manner:—

$$\begin{aligned} C_2 &= 12 \times 2 = 24 \\ H_5 &= 1 \times 5 = 5 \\ Cl &= 35\cdot 5 \\ & \overline{64\cdot 5} \end{aligned}$$

$$\text{Then } \frac{24 \times 100}{64\cdot 5} = 37\cdot 20$$

$$\frac{5 \times 100}{64\cdot 5} = 7\cdot 75$$

$$\frac{35\cdot 5 \times 100}{64\cdot 5} = 55\cdot 03$$

And these theoretical numbers are seen to be very close to those obtained by experiment.

	Theory.						Experiment.
C	37·20	.	.	.	.	.	37·20
H	7·75	.	.	.	.	.	7·90
Cl	55·05	.	.	.	.	.	54·95

26. What is the simplest formula you would assign to a substance containing—

Carbon	.	.	.	.	.	.	54·5
Hydrogen	.	.	.	.	.	.	9·2
Oxygen	.	.	.	.	.	.	36·3

in 100 parts?

Exercises on Section III. 157

27. Also to the following body :—

Carbon	88·20
Hydrogen	11·80
	100·00

28. And again to an organic base containing—

Carbon	63·78
Hydrogen	5·76
Nitrogen	3·32
Oxygen	27·14
	100·00

29. From the following percentages calculate formulæ for the several compounds :—

I.

Iron	60·5
Sulphur	39·5

II.

Iron	70
Oxygen	30

III.

Hydrogen	5·88
Oxygen	94·12

IV.

Sodium	32·79
Aluminium	13·02
Fluorine	54·19

V.

Carbon	39·31
Hydrogen	7·71
Oxygen	52·98

VI.

Carbon	68·67
Hydrogen	4·95
Oxygen	26·38

VII.

Carbon	49·05
Hydrogen	5·14
Nitrogen	28·61
Oxygen	17·20

VIII.

Carbon	42·00 ⎫
Hydrogen	6·46 ⎬
Oxygen	51·54 ⎭

IX.

Carbon	35·71 ⎫
Hydrogen	2·38 ⎪
Nitrogen	33·33 ⎬
Oxygen	28·58 ⎭

30. Find the formula of nitrosoterpene from these numbers:

Carbon	72·57 ⎫
Hydrogen	8·97 ⎬ in 100 parts,
Nitrogen	8·74 ⎭

and for nitrosoterpene hydrochloride from the following percentages:

Carbon	59·58
Hydrogen	8·07
Nitrogen	7·20
Chlorine	17·45

31. The silver salt of an organic acid was found by analysis to yield 47·1 per cent. of metallic silver. Determine its molecular weight.

In the formation of the silver salt from the acid, 108 parts of silver take the place of 1 part of hydrogen. Therefore,

      Molec. wt. of acid − 1 = molec. wt. of salt − 108

or   Molec. wt. of acid = molec. wt. of salt − 108 + 1.

In the example given 47·1 parts of silver are contained in 100 parts of the salt. So that 108 parts of silver are contained in $\frac{108 \times 100}{47\cdot1} = 229\cdot3$ parts of the salt. This is, therefore, a number identical with, or very near to, its molecular weight. The answer is, therefore,

      Molec. wt. required = 229·3 − 108 + 1 = 122·3.

Or, since the atomic weights of carbon, hydrogen, oxygen, and nitrogen are all integers, the fraction must be discarded, and the number becomes 122.

This corresponds with the formula of benzoic acid; verify it.

32. Aniline contains—

Carbon	77·4 ⎫
Hydrogen	7·5 ⎬ per cent.
Nitrogen	15·0 ⎭

Exercises on Section III.

and its platino-chloride contains 32·9 per cent. of platinum: to find its molecular weight and formula.

As explained in Chapter XV., p. 132, the platinum salts of nitrogenous bases are constituted on the same type as that of ammonia. Hence we may represent the formula of platino-chloride of aniline thus—

$$2(\text{Aniline} + HCl) + PtCl_4$$

or

$$\text{Aniline} + HCl + \frac{PtCl_4}{2}$$

The first question, then, is, what weight of platinum salt is represented by this formula? This is answered as follows:—

32·9 parts of platinum make 100 parts of platinum salt;

therefore 195 parts or one atom of platinum make $\frac{100 \times 195}{32 \cdot 9}$ parts, or 592.

We have now to subtract from this the platinum perchloride and hydrochloric acid; half the remainder is the molecular weight of the aniline.

$$592 - 337 - 73 = 182$$

and

$$\frac{182}{2} = 91$$

Now, taking the percentage composition of aniline, we have to calculate the proportions of the three elements contained in 91 parts of the base,

$$100 : 91 :: 77\cdot 4 : x$$
$$x = 70\cdot 4 \text{ carbon.}$$
$$100 : 91 :: 7\cdot 5 : y$$
$$y = 6\cdot 8 \text{ hydrogen.}$$
$$100 : 91 :: 15\cdot 0 : z$$
$$z = 13\cdot 6 \text{ nitrogen.}$$

Hence the formula is obtained by dividing these numbers by the respective atomic weights.

$$\frac{70\cdot 4}{12} = 5\cdot 86 \text{ atoms of carbon.}$$

$$\frac{6\cdot 8}{1} = 6\cdot 8 \text{ atoms of hydrogen.}$$

$$\frac{13\cdot 6}{14} = \cdot 97 \text{ atom of nitrogen.}$$

Hence, allowing for experimental error, which in this case is very small, the molecular formula required is

$$C\ \frac{5\cdot 86}{\cdot 97} \qquad H\ \frac{6\cdot 1}{\cdot 97} \qquad N\ \frac{\cdot 97}{\cdot 97} \quad \text{or} \quad C_6H_7N.$$

33. ·1442 gram of anthraflavic acid gave ·3712 gram of CO_2 and ·0448 gram of water. Calculate a formula.

34. A sulphide of tellurium and arsenic was analysed. ·6347 gram of the mineral gave ·2584 gram of tellurium, ·3978 gram of ammonio-magnesium arsenate ($MgNH_4AsO_4.H_2O$), and 1·6453 gram of barium sulphate. Calculate a formula.

35. The analysis of trichloracetyl urea gave the following results :—

(a) ·3210 gram gave ·2060 gram of CO_2 and ·0453 gram of H_2O ;
(b) ·0825 gram gave ·0109 gram of nitrogen ;
(c) ·1204 gram gave ·2510 gram of AgCl.

Calculate the formula of the compound.

36. Analysis of uranium pentachloride :—

Weight of substance taken	. . . .	·8955 gram
,, ,, U_3O_8 found	. . . .	·6038 ,,
,, ,, AgCl ,,	. . . .	1·4997 ,,

Calculate the formula.

37. ·3807 gram of benzoic acid gave ·9575 gram of CO_2 and ·1698 gram of water. And ·4287 gram of benzoate of silver gave ·2020 of silver. Calculate the rational formula of benzoic acid.

38. ·5828 gram of platino-chloride of caffeine left after ignition ·143 gram of platinum. What are the molecular weight and formula of caffeine, which contains

Carbon	. . . . . .	49·05
Hydrogen	. . . . . .	5·14
Nitrogen	. . . . . .	28·61
Oxygen	. . . . . .	17·20

per cent. ?

39. The platinum salt of a volatile organic base was found by analysis to have the following percentage composition :—Carbon, 9·5 ; hydrogen, 3·2 ; nitrogen, 5·7 ; chlorine, 42·0 ; and platinum, 39·0. The vapour density of the base was found to be 1·59 (air = 1). Calculate from these data its molecular formula.

40. The silver salt of an organic acid contained 62·44 per cent. of metallic silver. It also contains 17·34 per cent. of carbon and 1·73 per cent. of hydrogen. From these data endeavour to find a formula for the acid.

SECTION IV.

CHAPTER XVII.

CONDITIONS OF CHEMICAL CHANGE.

WE now proceed to discuss the general conditions under which chemical changes are brought about, together with some of the circumstances which have been found to modify the exercise of chemical attraction.

1. Bodies act upon each other chemically only when, according to the usual phrase, they are in absolute contact, that is to say, when they are so near to each other that the distance between them is immeasurably small.

In this respect chemical attraction differs from the attraction of gravitation, and of electrical and magnetic induction, all of which are capable of operating through distances which are appreciable by our senses and measurable by our instruments. On the other hand, it agrees thus far with *cohesion*, in virtue of which the molecules of bodies are held together in masses, and with *adhesion*, which causes surfaces of various kinds after being closely approximated to remain united.

2. Though in a few cases solids may be made to combine together by subjecting them to powerful pressure, or by continued trituration, union occurs only at their surfaces of contact, and cannot be completed unless one of the bodies, at least, is in the liquid or in the gaseous state. That is to say, the several parts of the acting masses must be free

to move, so that the mutual interpenetration required for chemical combination may be perfect.

3. Moderate elevation of temperature generally favours chemical combination, and is very frequently indispensable.

Thus it is a familiar fact that ordinary combustibles, wood, coal, gas, &c., do not begin to burn unless the temperature of some part of the mass is raised high enough. Again, iron and sulphur or copper and sulphur have no action on each other unless they are heated together very strongly. Copper has no action on sulphuric acid unless heated to about 130°, and in many cases chemical actions may be delayed or prevented altogether by keeping down the temperature.

On the other hand, a large number of chemical compounds can only be formed, and can only continue to exist, at moderately low temperature. Thus many salts unite with water, and ammonia unites with acids, to form definite compounds which are destroyed by heat. Many facts of this kind will be described in the chapter on 'Dissociation,' to which the reader is referred.

The production of polymeric compounds is a special case of combination occurring under the influence of moderate elevation of temperature.

Benzene, C_6H_6, for example, may be formed from acetylene, C_2H_2 (at a temperature below redness), and terpilene, $C_{10}H_{16}$, from isoprene, C_5H_8 (at about 280°). In some of these cases depolymerisation, a reversal of the process, occurs at a higher temperature.

4. The nature of the solvent or medium in which two bodies are presented to each other often influences their mutual action; and the physical character of the products, as to fusibility or volatility, frequently determines whether, and in what manner, chemical action may occur. Thus, when chloride of calcium and carbonate of ammonia are dissolved in separate portions of water, and the solutions then mixed together, a precipitate of carbonate of cal-

calcium is thrown down, and chloride of ammonium remains in the mother liquid. But if these two salts, calcium carbonate and ammonium chloride, are mixed together in the dry state and heated, a decomposition occurs which is the reverse of the last, and by which the original compounds, chloride of calcium and carbonate of ammonia, are re-generated.

Again, if acetic acid is poured into an aqueous solution of potassic carbonate, effervescence ensues from the escape of carbonic acid gas, and a solution of potassium acetate is formed. But if potassium acetate is dissolved in strong spirit of wine, a liquid in which carbonate of potassium is insoluble, a stream of carbonic acid gas transmitted through the solution is capable of decomposing the acetate, throwing down a precipitate of potassium carbonate.

We may accept it as very generally, though not quite universally, true, that when we mix together two soluble salts, which by double decomposition are capable of giving rise to an insoluble compound, that insoluble compound will be precipitated until complete decomposition of one or both the generating salts has taken place. If, for example, we take the two salts, barium chloride and sodium sulphate, knowing beforehand that barium sulphate is insoluble in water, we may safely assert that a precipitate will be formed when the aqueous solutions of these two salts are mixed together.

Similar observations hold good with regard to mixtures of compounds which, amongst them, contain the elements of a gas or body volatile at the temperature of the experiment. In many cases this volatile body is formed. Conversely, if we submit to pressure or to a very low temperature a mixture of substances which, under ordinary atmospheric conditions, evolves a gas, the chemical action is retarded, or sometimes prevented altogether.

5. When several bodies capable of acting on one another are mixed together, but the quantity of one of them largely preponderates over the rest, some curious results are fre-

quently brought about. When, for example, a solution of bismuth or antimony chloride in hydrochloric acid is mixed with a little water, no change is perceptible to the eye, but the addition of a larger quantity of water throws down a white precipitate of the oxychloride, BiOCl or SbOCl. These decompositions take place according to the following equations :—

$$BiCl_3 + OH_2 = BiOCl + 2HCl$$
$$SbCl_3 + OH_2 = SbOCl + 2HCl$$

That is to say, one molecule or 18 parts by weight of water decompose one molecule or 314·5 parts of bismuth chloride, or 226·5 parts by weight of antimony chloride ; and yet, if these proportions of the materials are brought into contact, only a partial decomposition takes place, and the reaction is completed only when a much larger quantity of water is added. In such a change as this we must remember that there are two antagonistic agents at work, namely, the water tending to decompose the chloride according to the equations given above, and the hydrochloric acid which is generated by that decomposition tending to reproduce the chloride. We may, therefore, consider that there are four bodies in presence of one another, and surrounded by water molecules.

Taking one case,

We have $BiCl_3$, OH_2, $BiOCl$, and HCl.

We may safely assume that when the number of water molecules present is augmented, the number of molecules of chloride decomposed by water in a given interval will increase, whilst the number of oxychloride molecules decomposed by the hydrochloric acid present will *pari passu* diminish, until the whole of the chloride is destroyed and precipitation is complete.

If now to the liquid holding the precipitate in suspension we add hydrochloric acid in excess the action is reversed, and the precipitate disappears again.

In such experiments as these the decomposition seems to proceed continuously as the quantity of the acting body is increased, any alteration in the proportions, however small, apparently producing a corresponding alteration in the extent to which decomposition takes place.

Similar actions and reactions must be supposed to ensue when salts in solution are mixed together, or when acids are mixed with salts, with alcohols, or with ethers, and very generally when several substances meet together, the final condition of equilibrium being attained only after a lapse of time which in some cases is long enough to be measurable, and the quantities of the several products varying according to the relative masses of the active ingredients and according to the relative intensity of their affinities.

6. The action of two substances upon each other is often promoted by the presence of a third substance which has the power of combining with one of the products of change. In the 'basic' processes for the production of steel, pig-iron containing phosphorus is melted in a vessel lined with calcined dolomite, and some common lime is generally added, whilst air is either blown through the molten metal or over the surface. Under these circumstances the oxidised phosphorus unites with the lime, forming a phosphate, which remains undecomposed by contact with the iron. Whereas if the lining consists of an 'acid' substance like silica, as in the original form of the Bessemer process, phosphate of iron is formed, which is immediately reduced by the heated iron, and the phosphorus is thus returned into the metal, and cannot be removed.

A somewhat similar action may be supposed to occur in the production of sodium sulphate by Hargreaves' process. Sulphur dioxide, oxygen, and water vapour react upon one another very slowly to produce sulphuric acid, but when these gases are brought into contact with common salt at about 400° sodium sulphate and hydrochloric acid are freely produced.

Another example of the action of a third substance is afforded by the common methods for producing iodo-substitution compounds. (See p. 222.)

In each of the foregoing cases the explanation of the effect is tolerably obvious. There are, however, a great many similar changes, the *rationale* of which has not yet been made out.

Deacon's process for making chlorine—viz. by passing a mixture of air and hydrochloric acid over a heated copper salt—is one of these. So also is the action of a small quantity of acid in promoting the combination of turpentine with water to form terpin—

$$C_{10}H_{16} + 2H_2O = C_{10}H_{20}O_2;$$

the hydrolysis of such compounds as cane sugar—

$$C_{12}H_{22}O_{11} + H_2O = \underset{\text{Dextrose.}}{C_6H_{12}O_6} + \underset{\text{Lævulose.}}{C_6H_{12}O_6};$$

and the elimination of the nitrogen of cyanides and isocyanides—

$$HCN + H_2O = \underset{\text{Formic acid.}}{CH_2O_2} + \underset{\text{Ammonia.}}{NH_3}.$$

Another example is afforded by the application of aluminium chloride in promoting union of hydrocarbon radicles; as, for, instance, when added to a mixture of benzene and ethyl chloride—

$$C_6H_6 + C_2H_5Cl = HCl + \underset{\text{Ethyl benzene.}}{C_6H_5.C_2H_5}$$

In all these cases there is strong probability in favour of the hypothesis that the action proceeds through a cycle, an unstable compound being formed in small quantities at a time and subsequently decomposed, the 'catalyst,' as the active substance may be called, returning finally to the same state of combination as at first.

Certain decompositions, which proceed slowly under the action of heat alone, are also effected more quickly or at a lower temperature in the presence of a small quantity of a second substance which apparently remains unchanged at the end of the process. The most familiar example of this is the decomposition of potassium chlorate by heat in the presence of a small quantity of manganese dioxide, ferric oxide, or cupric oxide, also the evolution of oxygen from calcium hypochlorite by the action of a little oxide of cobalt.

In such cases as these, again, there is doubtless alternate oxidation and deoxidation of the one substance by the other.

But perhaps the most remarkable examples of the action of a minute quantity of a third substance upon the progress of action between two others are found in the observations which have been made within the last few years upon the combustion of certain gases in oxygen.

It has been found by Professor Dixon that a mixture of dry carbonic oxide and oxygen does not explode when an electric spark is passed through it, but the addition of a minute trace of water vapour makes the mixture explosive, and the rapidity of the explosion increases with the quantity of steam introduced. The presence of a minute quantity of various gases which contain hydrogen, such as hydrogen sulphide and chloride, ether vapour, or pentane, has a similar effect; but the addition of dry nitrous oxide or carbon disulphide was not found to render the mixture inflammable.

Why steam and hydrogen compounds, which by interaction with oxygen produce steam, should be necessary for the propagation of the flame through the gas has not been fully explained. Probably at the temperature of the spark the water molecule is less stable than the oxygen molecule, and the decomposition expressed by equation I occurs.

$$\text{I.} \quad CO + H_2O = CO_2 + H_2$$

$$\text{II.} \quad \left\{ \begin{array}{c} H_2 + O \\ CO + O \end{array} \right\} = H_2O + CO_2$$

The second phase may be an example of the effect of two attractions acting concurrently, the hydrogen and carbonic oxide in equation **II** dividing a molecule of oxygen between them, and the H_2O then reacting with another molecule of CO, and so alternately.

7. The peculiar activity of substances in the 'nascent state' has been already described (Chap. XV., p. 126), as well as some other examples of concurrent chemical attractions.

CHAPTER XVIII.

DISSOCIATION.

PEROXIDE of nitrogen is a compound which exhibits very remarkable phenomena. Below 10° it forms colourless crystals which melt at higher temperatures, forming a yellow liquid. This liquid boils at about 22°, though, like water, it evaporates at temperatures below this. The vapour is yellow, but when gradually heated it first deepens in colour, till at 100°, and above, it becomes dark orange-red. If the temperature is still further raised to 400°—500° it suddenly becomes colourless. On cooling these changes proceed in reverse order, the dark-coloured gas being first formed, then the pale yellow gas, ultimately condensing to a pale yellow liquid. At the lowest temperature at which the density of the vapour has been determined, namely, at 4·2°, it was found to be 2·588 (air = 1), whilst the formula N_2O_4 requires the vapour density to be 3·178. At about 150° the density was found to be 1·58, which closely agrees with the value 1·59 corresponding to the formula NO_2. The loss of colour at higher temperatures is doubtless due to the formation of nitric oxide, NO, and oxygen, both colourless gases, though the vapour density in this condition has not been ascertained.

A lump of common 'sesquicarbonate' of ammonia, or

a solution of the same salt, exposed to atmospheric air at common temperatures, loses ammonia, recognisable by its pungency, and carbon dioxide, till ultimately no residue remains. Or when a solution of ammonium oxalate, chloride, or nitrate is boiled, a similar escape of ammonia may be observed, and the liquid acquires a distinctly acid reaction.

$$(NH_3)_2H_2C_2O_4 \underset{\text{Ammonium oxalate.}}{} = \underset{\text{Ammonium acid oxalate.}}{NH_3H_2C_2O_4} + NH_3$$

And if in either of these cases the ammonia thus evolved is brought into contact at a lower temperature with the solution from which it was produced, it will enter again into combination, and the original compound will be regenerated.

Again, when calcium carbonate is heated strongly in a vessel from which the air has been more or less completely removed by the air-pump, it suffers decomposition into lime, which remains behind, and carbon dioxide gas, which fills the vessel, $CaCO_3 = CaO + CO_2$; and this decomposition proceeds until the evolved gas acquires a certain density or tension, which increases as the temperature rises. If now the whole is allowed to cool, the carbonic acid gas slowly recombines with the lime, and a vacuum is once more established. Calcium carbonate splits up in the same manner when a current of steam, or other gas into which the carbonic anhydride can diffuse, is passed over it whilst under the action of heat.

Decompositions like that of nitric peroxide gas, of the ammonium salts, or of calcium carbonate under the influence of heat, are examples of what is known as *dissociation*, or, as it is sometimes more precisely termed, *thermolysis*. The word dissociation has acquired by careless use some degree of ambiguity, but in these pages its application will be restricted solely to those cases of decomposition in which certain bodies are resolved at an elevated temperature into simpler bodies, which are capable of reuniting and reproducing the original compound when the temperature is again allowed to fall.[1]

We shall now consider several other cases of dissociation.

A crystal of sulphate of copper remains unaltered at ordinary temperatures in moist air; but if moderately heated, it loses its water of crystallisation, and crumbles down to a white powder.

The same kind of decomposition takes place in many cases at the temperature of the air, and under ordinary conditions. Salts, which thus readily part with their water of crystallisation and fall away to powder, are said to be *efflorescent*. Sulphate, carbonate, and phosphate of sodium afford examples of this kind of dissociation, which, however, presents nothing remarkable beyond the fact of occurring at comparatively low temperatures. In this respect, however, these phenomena are surpassed by those exhibited by salts, which, under ordinary circumstances, are deposited from solution void of water of crystallisation. As already mentioned (Chap. II., p. 15), such compounds have been found in every instance to combine with water when crystallised at temperatures below zero. The explanation of the existence of anhydrous crystals is simply that dissociation of the salt from the water occurs at or below the temperature at which the crystals are deposited. And even those salts which usually combine with water of crystallisation may be obtained in a lower state of hydration, or altogether destitute of water, by causing them to crystallise at more or less elevated temperatures. Sodium sulphate furnishes a case in point. The crystals of this salt formed at ordinary temperatures contain ten molecules of water of crystallisation combined with one of the salt ($Na_2SO_4.10OH_2$); when formed at 18° they contain seven molecules of water ($Na_2SO_4.7OH_2$), whilst a solution heated to 34° yields crystals which contain

---

[1] The decomposition of ferric salts and other similar compounds, when heated with water, ought not to be represented as cases of dissociation, being in reality the effects of mass. (See Chap. XVII.)

Abnormal Vapour Densities.

no water at all. Similar phenomena are beautifully exhibited when the solutions of some coloured salts are heated. Chloride of cobalt especially lends itself to this kind of reaction. This salt forms red crystals consisting of $CoCl_2 + 6OH_2$, and when dissolved in water it gives a red solution; but if the temperature is raised even very slightly the liquid changes in colour, becoming successively purple and blue, and these changes probably correspond with the formation of increasing quantities of the compounds $CoCl_2 + 4OH_2$ and $CoCl_2 + 2OH_2$ respectively. When the temperature is again allowed to fall, the solution reassumes its original red colour.

In the early attempts to apply the law of Avogadro to the determination of molecular weights, many cases of 'abnormal vapour density' presented themselves and gave rise to much controversy. Thus sulphuric acid is entirely vaporisable at a moderate temperature, but the relative density of the vapour of this compound is only about one-fourth, and not one-half, of the molecular weight as represented by the formula H_2SO_4, which accords with the bibasic character of sulphuric acid. In other words, vapour of sulphuric acid instead of being, in accordance with the general rule, forty-nine times as heavy as an equal volume of hydrogen taken at the same temperature, is only $\frac{49}{2} = 24\frac{1}{2}$ times as heavy or thereabouts.

Ammonium chloride, phosphorus pentachloride, and calomel afford instances of the same kind. These anomalies are explained by the hypothesis that when these compounds are volatilised they are at the same time resolved by dissociation into two distinct products; sulphuric acid into sulphuric anhydride and water, ammonium chloride into ammonia and hydrochloric acid, phosphorus pentachloride into phosphorus trichloride and chlorine, calomel into mercuric chloride and mercury. And in these particular cases direct experimental evidence can be adduced that these products exist in the several vapours.

In the case of sulphuric acid and of ammonium chloride advantage has been taken of the difference in the diffusibility of the products into which these bodies are supposed to dissociate. Thus the vapour of sulphuric anhydride is much heavier and consequently less diffusible than vapour of water, so that when sulphuric acid is heated for several hours in a vessel with a capillary orifice, the water vapour escapes more rapidly than the sulphuric anhydride, and the latter gradually accumulates in the residue. Pentachloride of phosphorus has also been resolved into free chlorine and phosphorus trichloride, which may be to some extent separated by diffusion.

In the case of sublimed calomel we can appeal to the fact, well known to manufacturers, that it invariably contains small quantities of corrosive sublimate, and of metallic mercury, which have escaped recombination during the cooling down of the vapour.

Direct proof that dissociation has occurred is, however, not possible in all cases, though there is usually some collateral evidence available in support of the natural inference from the vapour density. An interesting case is presented by the halogens, especially iodine. The vapour density of this element at temperatures up to about 700° is 8·8 (air = 1); above 700° the density diminishes rapidly, being reduced at 1500° (the pressure being ·3 atmosphere) to nearly half this, or 4·6. The conclusion seems to be inevitable that at the lower temperature molecules of iodine consist of two atoms, I_2, whilst at the higher they dissociate into separate atoms.

The phenomena of dissociation appear to result from two opposite and reciprocal actions—the one of decomposition, the other of recombination proceeding simultaneously At low temperatures, when the composition of hydric sulphate, for example, is represented by the formula H_2SO_4 or

H_2O, SO_3, the number of molecules decomposed, into H_2O and SO_3, is exactly counterbalanced by the number of molecules which are reconstituted in the same period by the reunion of these two substances. As the temperature rises and the agitation of the molecules in the mass becomes more vigorous, the number of molecules which undergo decomposition progressively increases, whilst the recombination *pari passu* continually decreases, till at length a point is reached in which the SO_3 molecules and the OH_2 molecules become indifferent to each other, and recombination no longer takes place. When the temperature is allowed to fall these processes are reversed, and for every degree of temperature a certain definite relation subsists between the decomposition and recomposition, so that a kind of equilibrium is maintained. The compensating process being at high temperatures annulled, decomposition is complete; the vapour consists throughout of a uniform mixture of two different kinds of molecules, and consequently, by the law of Avogadro, it occupies twice the volume it would otherwise fill if dissociation did not take place.

The rate and extent of dissociation in a given case therefore, independently of the chemical character of the substance concerned, depends upon at least three conditions; namely, (1) temperature, (2) pressure, and (3) the proportion of the products of dissociation to the undecomposed substance. If one such substance is introduced at starting, the dissociation may be much impeded or even prevented altogether. Thus if phosphorus pentachloride is mixed with a sufficient quantity of the trichloride, PCl_3, it gives a normal vapour density corresponding to the formula PCl_5.

It is obvious from what has gone before that a considerable number of compounds can only exist in the liquid or solid state, and never, so far as known, in the state of gas. A parallel has been traced between the phenomena of dissociation and evaporation, or, in other words, between the separation of molecules and atoms of different kinds and the separation of molecules of the same kind in passing under the influence of heat or diminished pressure from the solid or liquid to the gaseous state. Thus the vaporisation of water from a crystallised salt seems to be a phenomenon of the same order as the escape of water vapour from ice. Further, the progress of the change from liquid acetic acid (see next page) to the same in a perfectly gaseous state at 250° is gradual and continuous throughout, as also in such a decomposition as that of calcium carbonate under the influence of heat. In all such cases the rate of vaporisation is governed by temperature and pressure, and is attended by the absorption of the energy of heat, and reversal of conditions leads to a restoration of the original state of the substance.

One remarkable instance of the formation of a chemical compound by compression of a mixture of two gases, and its destruction by diminishing the pressure whilst the temperature remains unaltered, is afforded by phosphonium chloride. When equal volumes of PH_3 and HCl are mixed at 14° no visible combination occurs, but on applying a pressure of twenty atmospheres white crystals are formed of PH_4Cl, which disappear into gas when the pressure is reduced.

Among the cases of dissociation which are commonly recognised it is advisable to distinguish those in which a double decomposition must be supposed to occur between two molecules from those in which the products are derived from a single molecule. Thus in such examples as the following there is evidently an exchange of constituents between two molecules of the same kind:—

$$HI + HI = H_2 + I_2$$
$$H_2O + H_2O = 2H_2 + O_2$$
$$CO_2 + CO_2 = 2CO + O_2$$
$$SCl_2 + SCl_2 = S_2Cl_2 + Cl_2$$

The following list includes some of the most important cases of simple dissociation without double decomposition.

Examples of Dissociation.

Molecule of substance before dissociation	Products of complete dissociation at $t°$	$t°$ approximately. Pressure 1 atmos.
I_2 vapour below 700°	$I + I$	1500°
Br_2 vapour	$Br + Br$	1500°
Cl_2 gas	$Cl + Cl$	above 1600°
S_6 vapour at 482°	$S_2 + S_2 + S_2$	940°
N_2O_4 vapour below 0°	$NO_2 + NO_2$	160°
$[N_2O_3]_n$ liquid	$NO_2 + NO$ vapour	all temps.
PCl_5, b.p. 160°–165°	$PCl_3 + Cl_2$	300°
$SbCl_5$	$SbCl_3 + Cl_2$	
$[Hg_2Cl_2]_n$ solid	$HgCl_2 + Hg$ vapour	950°
$[Al_2Cl_6]_n$ solid	$AlCl_3 + AlCl_3$ vapour	800°
$[ICl_3]_n$ solid	$ICl + Cl_2$ vapour	all temps.
H_2F_2 vapour at 20°	$HF + HF$	
$[NH_4Cl]_n$ solid	$NH_3 + HCl$	350° to 940°
$[NH_4Br]_n$ solid	$NH_3 + HBr$	440° to 772°
$[NH_4I]_n$ solid	$NH_3 + HI$	440° to 772°
$[PH_4I]_n$ solid	$PH_3 + HI$	
$[(C_2H_5)_4NI]_n$ solid (tetrethylammonium iodide)	$(C_2H_5)_3N + C_2H_5I$ vapour (triethylamine + ethyl-iodide)	all temps.
$[NH_4HS]_n$ solid	$NH_3 + H_2S$	57°
$[H_2SO_4]_n$ liquid	$H_2O + SO_3$	416°
$[C_5H_{11}Cl]$ liquid (isoamyl chloride)	$C_5H_{10} + HCl$ (amylene)	300°
$[SCl_4]_n$ liquid at $-22°$	$SCl_2 + Cl_2$	8°
$[C_2H_3Cl_3O_2]_n$ solid, boils 95° (chloral hydrate)	$C_2HCl_3O + H_2O$ (chloral)	100°
$[C_2H_4(C_2H_5)Cl_3O_2]$ solid (chloral-alcoholate)	$C_2HCl_3O + C_2H_6O$ (chloral + alcohol)	116°
$[C_2H_4O_4 + CH_2O_2] = 4$ vols. (formic acid vapour at b.p. 101°)	$CH_2O_2 + CH_2O_2 + CH_2O_2$ (formic acid)	220°
$[C_4H_8O_4 + C_2H_4O_2] = 4$ vols. (acetic acid vapour at b.p. 119°)	$C_2H_4O_2 + C_2H_4O_2 + C_2H_4O_2$ (acetic acid)	250°
Solid compounds of salts with ammonia, e.g. $[AgCl.3NH_3]_n$	Salt + $n NH_3$	
Salts united with water of crystallisation	Anhydrous salt + $n H_2O$	

CHAPTER XIX.

THERMAL AND ELECTRICAL EFFECTS OF CHEMICAL ACTION.

EVERY chemical change, whether it occurs in one substance alone or as a result of the contact of two or more substances, is attended by a redistribution of the total energy of the system. This is most commonly manifested in the form of absorption or evolution of heat. There is, however, nothing in this to distinguish chemical change from physical or mechanical changes, such as fusion, vaporisation, or alteration of density.

Suppose a mass of ice at 0° to be so gradually heated that it melts, but the resulting water remains at the same temperature. The amount of energy in the form of heat thus supplied would raise the temperature of the same mass of water from 0° to 80°. If the mass of ice is a gram we say that 80 thermal units, or 80 units of heat, or 80 calories, are required for its fusion.[1] In similar terms 100 units are required to raise the temperature of a gram of water from 0° to the boiling-point under a pressure of one atmosphere, and 537 units more to change it into steam. At a much higher, unknown, temperature the disruption of one gram of water gas into a mixture of hydrogen and oxygen would require at least 3,780 units, and probably much more. All these changes are reversible, and when reversed the same quantities of heat are successively given out.

The more obviously chemical changes which have been studied in relation to the thermal changes which attend them may be ranged under the following heads :—

1. Combinations and decompositions at low tempera-

---

[1] Sometimes the unit of mass referred to is a kilogram. The corresponding thermal unit may be called a large calorie.

tures. 2, Combinations at high temperatures (combustion). 3. Dissolution in liquids. 4. Allotropic and isomeric changes.

It will be convenient to consider these separately, and in the present chapter to attend chiefly to the first division. The others will be referred to in their proper places. (See Chaps. III., XX., XXI.)

For the forms of calorimeters, the methods of experiment, and the calculation of results, books on physics must be consulted.[1]

The estimation of the heat evolved during the neutralisation of an acid by a basic hydrate is perhaps one of the least difficult of performance, and most intelligible among thermo-chemical measurements. When solution of hydrochloric acid, for example, is mixed with solution of soda sufficient to neutralise it heat is evolved, the amount given out, however, varying to an appreciable extent according to circumstances, whereof the most important are (1) the proportion of water present, and (2) the temperature of the solutions when mixed.

It is obvious, therefore, that the heat produced results from a number of changes, part of which may be physical or mechanical, and part chemical, though in these cases, at any rate, there is no means of distinguishing them. Hence the true thermal result of such a reaction as

$$HCl + NaHO = NaCl + H_2O$$

is unknown. After all, however, the practical object is to compare together the action of substances of the same kind under circumstances as nearly alike as possible.

Thomsen obtained the following values for the heat of neutralisation of the common acids. This expression means the number of calories evolved when one molecule of a monobasic acid, or an equivalent quantity of polybasic acid, is mixed with one molecule of sodium hydroxide, both

---

[1] Experimental details and results will be found in Muir's *Thermal Chemistry*.

being previously dissolved in 100 molecules of water and at the temperature of 16° to 18° C.

Heats of Neutralisation of Acids by Soda.

	Calories
HCl	13,740
HBr	13,748
HI	13,721
HF	16,272
HNO_3	13,617
H_3PO_2	15,160
HPO_3	14,510
$\frac{1}{2}H_2SO_4$	15,690
$\frac{1}{2}H_3PO_3$	14,228
$\frac{1}{3}H_3PO_4$	11,343
$HCHO_2$	13,450
$HC_2H_3O_2$	13,400
$HC_2H_2ClO_2$	14,280
$HC_2HCl_2O_2$	14,830
$HC_2Cl_3O_2$	13,920
$\frac{1}{2}H_2C_2O_4$	13,840
$\frac{1}{2}H_2C_4H_4O_4$	12,400

The heat of neutralisation of the majority of acids examined ranges between about 10,000 cal. and 15,000 cal. It will be noticed that hydrofluoric acid stands at the top of the list; that hydrochloric, hydrobromic, and hydriodic acids are nearly equal to one another. It is also interesting to notice the effect on the thermal equivalent of introducing chlorine in place of hydrogen into acetic acid.

When a molecule of an acid in aqueous solution reacts upon sodium hydroxide, the thermal change is nearly, but not exactly, proportional to the quantity of soda measured as 1, 2, 3, or 4 molecules, according as the acid is mono-, di-, tri-, or tetra-basic. In the neutralisation of dibasic acids, such as the following, for example, it sometimes happens that the evolution of heat is greater on addition of the first molecule of soda, sometimes on addition of the second. Thus on neutralising one molecule of sulphuric acid by soda the first molecule of NaHO added gives 14,750 C.; the second molecule gives 16,630 C. (mean 15,690 C.). With sulphurous acid, on the other hand, the first stage towards neutralisation gives 15,870 C., the second 13,100 C. (mean 14,485 C.). What the explanation of this difference may be is not known. It has been supposed to support the idea that the constitution of the two acids may be different, sulphurous acid and others such as carbonic acid, which behave in a similar manner, being represented as monobasic. It is conceivable, though not very probable, that sulphurous acid may be

$$(SO_2) \begin{cases} OH \\ H \end{cases} \text{ or } \begin{matrix} HO-S-H \\ \diagup \diagdown \\ O \quad O \end{matrix}$$

A similar formula for carbonic acid cannot, however, be admitted.

A substance in a chemically free or unsaturated condition may be compared to a body which has been raised to a height. In order to get it into such a position, energy in some form must be expended, but the whole of that energy is recoverable in the form of heat or mechanical effect by the descent of the body to its former level. So, when a chemical compound is resolved into its constituents, mechanical work, or rather its equivalent in the form of the kinetic energy of heat or electricity, is transformed into the potential energy of chemical separation; but when the elements come together again in their original order and proportion the same amount of energy is again available, chiefly in the form of heat.

Whenever much heat is evolved, it is tolerably safe to conclude that the resulting compound is a very stable one. Thus when 1 gram of hydrogen combines with $35\frac{1}{2}$ grams of chlorine the result is not only the production of $36\frac{1}{2}$ grams of hydrochloric acid gas, but a certain amount of kinetic energy which becomes manifest in the form of heat.

The amount of heat is represented by 22,000 units. In order to reverse this effect, and bring about the decomposition of the compound gas into its constituents, and restore them to their original physical condition, precisely the same amount of energy must be supplied.

Further, all chemical changes which result in the production of more stable from less stable forms of matter, whether by way of combination or of so-called decomposition, are attended by the evolution of heat; and, as matter cannot pass into conditions of instability without the aid of external forces, all spontaneous chemical changes tend towards the production of that body or system of bodies, in the formation of which the greatest amount of heat is given out. This has been called the 'principle of maximum work.' Such, for example, are the action of acids upon bases in solution, the displacement of the halogens by one another in their compounds, the displacement of metals by one another in their salts, &c.

From the following examples it will be seen that, with the thermal values as data, we can predict what will happen if one of the metals is placed in a solution of the chloride of one of the others, which of the oxides is most easily reducible by the action of carbon or other agent, why chlorine decomposes bromides and iodides, and so on.

Heat evolved in the Formation of One Formula-weight in Grams.

Chlorides in Watery Solution.		Hydrated Oxides, Solid.	
$MgCl_2$	186,930	MgO,H_2O	148,960
$ZnCl_2$	112,840	ZnO,H_2O	82,680
$CdCl_2$	96,250	CdO,H_2O	65,680
$HgCl_2$	59,860	—	—
$FeCl_2$	99,950	FeO,H_2O	68,280
$CuCl_2$	62,710	$CuOH_2O$	37,520
$\tfrac{2}{3}AuCl_3$	18,174	—	—

Chlorides.		Bromides.		Iodides.	
KCl	101,170	KBr	90,230	KI	75,020
NaCl	96,510	NaBr	85,580	NaI	70,300
$\frac{1}{2}CaCl_2$	93,820	$\frac{1}{2}CaBr_2$	82,889	$\frac{1}{2}CaI_2$	67,670
HCl	39,320	HBr	28,380	HI	13,170

In other cases, however, where the products of the change remain in the field of action the change may never be completed, although it might be expected from the several thermo-chemical values taken separately. Thus, when an acid is added to the aqueous solution of a salt of another kind, it by no means follows that decomposition will be complete, even though the heat of neutralisation of the acid added is much larger than that of the acid formed by its action in the solution. Thus sulphuric acid added in equivalent quantity to a carbonate causes its complete decomposition, but added to a solution of a chloride the salt is completely changed into a sulphate only when the hydrochloric acid is removed in proportion as it is formed. (See also Chap. XVII., pp. 162 to 165.)

In general, then, products of chemical combination contain less energy than the materials from which they are formed, the difference being disposed of in the form of heat. But in some few cases the reverse is true ; that is, the compound contains more energy than its constituents in the state in which they are usually known. Thus the formation of hydriodic acid, of nitric oxide, of ozone, and probably of acetylene and carbon bisulphide, is attended by absorption of heat, which is given out if these substances are decomposed.

Comparing together the thermal effects of causing one gram of hydrogen to combine with equivalent quantities of chlorine, bromine, and iodine in presence of water, heat is in each case evolved.

$$H + Cl + 400\ H_2O \text{ gives } 39,320 \text{ calories}$$
$$H + Br + 400\ H_2O \quad ,, \quad 28,380 \quad ,,$$
$$H + I + 400\ H_2O \quad ,, \quad 13,170 \quad ,,$$

A large part of the effect is, however, due to the solution of the resulting hydracids in the water, and if the value of this is subtracted we get a different result.

	Heat of formation in presence of water		Heat of solution		Heat of formation of gas
HCl	39,320	—	17,320	=	22,000
HBr	28,380	—	19,940	=	8,440
HI	13,170	—	19,210	=	− 6,040

The negative sign means that in production of hydriodic acid from hydrogen and iodine gases *absorption* of 6,040 units occurs for every unit of hydrogen which combines. The decomposition of hydriodic acid gas is attended by evolution of heat. We know how prone this compound is to decomposition, and that even a solution of it, provided it is strong, forms a powerful reducing agent, iodine being set free.

In very dilute solutions, however, its reducing powers are much less, as may be inferred from the thermal value of its formation in presence of much water.

But in such cases as hydriodic acid, after all due allowances have been made for the heat which becomes latent in consequence of the change from solid to liquid, or gas, the actual absorption of heat which is observed must be attributed to the fact that energy is used up in the decomposition of the elementary molecules which precedes, or is simultaneous with, the formation of the molecules of the compound. Thus, when hydrogen combines with iodine, the change is not simply

$$H + I = HI$$

but is, strictly speaking, a double decomposition; thus—

$$\left.\begin{matrix}H\\H\end{matrix}\right\} + \left.\begin{matrix}I\\I\end{matrix}\right\} = \left.\begin{matrix}H\\I\end{matrix}\right\} + \left.\begin{matrix}H\\I\end{matrix}\right\}$$

Now, if the heat required to effect the separation of H from H, and I from I, is greater than the amount of heat generated when 2H combines with 2I, then the result will be negative ; no heat will be given out, but a certain amount must be supplied from external sources.

From this consideration it is apparent that even in those cases of chemical action in which an evolution of heat is actually observed, this heat represents only the surplus energy remaining over after the breaking up of the molecules of the original substances employed in the experiment.

Not only is heat generated when chemical action occurs, but in at least a great number of cases a definite amount of electricity is developed. Imagine a plate of zinc plunged into a solution of hydrochloric acid. The chlorine unites with the zinc, and leaves the hydrogen to escape in bubbles from the surface of the metal. If now a plate of platinum, which is not acted upon by hydrochloric acid, is immersed in the same liquid, and connected with the zinc by a wire, it will be observed that the hydrogen is no longer disengaged from the zinc plate, but collects upon the platinum in bubbles which rise to the surface of the liquid. Such a combination of two metallic plates immersed in a liquid and connected together outside constitutes a voltaic cell. A number of these cells joined together forms a battery. If the wire is cut in two, that part which proceeds from the zinc can be shown, by a gold-leaf electroscope, to be charged with negative electricity, whilst the wire from the platinum is positive. If these wires are then attached to platinum plates dipping into a solution of some salt, such as iodide of potassium, which is easily decomposable, the elements of the salt are separated, the metal going to the negative electrode or terminal connected with the zinc of the battery, and the non-metallic element making its appearance at the electrode belonging to the platinum plate [1] of the battery, that is at the positive pole. This process of decomposition

---

[1] The elements which make their appearance at the negative pole are often referred to as *electro-positive*, whilst those which collect at the positive pole are *electro-negative*. The metals are generally electro-positive, the non-metals electro-negative.

by the voltaic current is called 'electrolysis.' It deserves to be noticed that it can only take place when the body to be operated upon is in the liquid state. Now, just as the amount of heat evolved by a given chemical combination is definite and constant under the same circumstances, so the electrical effect produced by contact of the same bodies under the same circumstances is constant. When electrolysis occurs in the body through which the current is passing, the quantities of its constituents which are liberated are always proportional to their chemical equivalents, and, provided secondary actions are guarded against, these quantities are also chemically equivalent to the materials consumed in each cell of the battery. If, for example, in each cell 32·5 grams of zinc are dissolved, the current passing simultaneously through solutions of iodide of potassium, bromide of potassium, acidulated water, and sulphate of copper would set free in the

1st cell, 127 grams of iodine, 39·1 grams of potassium.[1]
2nd „ 80 „ bromine, 39·1 „ „
3rd „ 8 „ oxygen, 1 „ hydrogen
4th „ 8 „ oxygen, 31·75 „ copper

The power of effecting electrolytic decomposition exhibited by different combinations of metals and liquids depends entirely upon the chemical characters of those bodies. The more energetic the chemical action the more powerful will be the electro-motive force of the combination.

As to the *rationale* of the decomposition effected by the current opinions differ;[2] certain phenomena of electrolysis seem to be at present inexplicable. Thus liquid carbon dioxide, cyanogen, carbon disulphide, hydrochloric acid, and stannic chloride, when perfectly free from water, do not

[1] Hydrogen and caustic potash would, of course, be the actual products at the negative electrode.
[2] See Chap. XXII.

conduct and are not decomposed, and even water itself when quite free from dissolved salts seems to be in the same predicament. On the other hand, certain salts, such as the chlorides and iodides of lithium, calcium, barium, lead, and silver, when melted, not only conduct but undergo electrolytic decomposition very readily.

It is not easy to see in what respect melted silver chloride differs in chemical constitution from hydrogen chloride or stannic chloride, yet the former yields, by the action of a current from one or two Grove's cells, metallic silver and chlorine respectively at the two electrodes; whilst the latter not only offer great resistance, but show no signs of decomposition when the most powerful current yet obtained is passed through them. But the case is altogether different if water is added. Aqueous solutions of hydrochloric acid or of tin chloride are decomposed with the utmost facility by very feeble currents. When solutions are electrolysed both the dissolved substance and the water seem to undergo decomposition, the relative proportions in which the products make their appearance depending on the strength of the solution, and apparently also on the strength of the current. Thus a concentrated solution of hydrochloric acid yields only hydrogen and chlorine; a weak solution yields only hydrogen and oxygen. In this latter case it may be that chlorine is really liberated, but has time or opportunity, when there are many water molecules around, to attack some of them, reproduce hydrochloric acid, and liberate the oxygen. Possibly, however, since water alone and hydrochloric acid alone resist the current, the real electrolyte may be some compound of the two existing in the solution. (See also Chap. XXII., p. 205.) From solution of copper sulphate, copper is alone deposited at the negative pole when the current is weak, hydrogen also making its appearance if the current used is strong.

CHAPTER XX.

COMBUSTION.

THE burning of wood, coal, charcoal, and other matters commonly employed as fuel is a process with which every one is familiar. Chemists know that the production of fire in the usual way is attended not merely by the consumption or alteration of the fuel, but by changes in the surrounding atmosphere, and that the presence of a sufficient supply of air is an indispensable condition in the operation. They explain the phenomena by stating that the process of burning consists essentially in the combination of the elements of the combustible body with the oxygen in the air, so much heat being developed that more or less of the solid combustible and of the products of combustion are raised to such a temperature that they emit light. Notwithstanding, then, that in ordinary fires the coals disappear and seem to be destroyed, they do in reality only evaporate away in the form of carbon dioxide and water; and if these products could be collected and weighed, their weight would be found to be made up of the united weight of the carbon and hydrogen of the coal, and the oxygen which is taken from the air.

The phenomena of combustion may be observed equally well when other materials are employed. Thus copper burns in vapour of sulphur, hydrogen will burn in chlorine, whilst phosphorus and several metals become ignited spontaneously when introduced in the proper condition into the same gas. In every such case the resulting product consists of a compound of the body which is burned with one or other of the constituents of the gaseous atmosphere which surrounds it. From this it is evident that the terms 'combustible' and 'supporter of combustion,' as generally employed, involve an error if they are taken to imply any difference of function; for that which in one experiment occupies the position of combustible may be made the supporter of combustion or atmosphere in another. It is easy to show, for example, that not only will a jet of hydrogen burn in oxygen gas, but that a jet of oxygen burns equally well when surrounded by hydrogen.

Previously to the discovery of oxygen by Priestley, and the establishment of the modern theory of combustion by Lavoisier at the close of the last century, a remarkable theory had been for upwards of fifty years adopted by chemists. This was the celebrated theory of phlogiston,[1] proposed by Stahl.[2] This phlogiston was supposed to be a substance of great tenuity, which, by combining with incombustible bodies, rendered them combustible. When such bodies are burnt, it was imagined that the escape of the phlogiston in a peculiar condition of vibratory motion gave rise to the phenomena of fire. At the time the idea was originally introduced little was known of the part which the air plays in all ordinary burning. When accumulated facts proved conclusively that bodies by burning increase in weight, some attempts were made to prop up the theory by assuming that the presence of phlogiston gave bodies lightness instead of weight. The merit of the idea, however, lay not so much in providing an explanation of certain special cases of combustion, as in referring all cases of burning to a common cause, and in showing that the property of combustibility is capable of being transferred from one body to another. Oxides of the metals, for example, were regarded as ashes, or 'calces,' of the metal left after the escape of their phlogiston, which could be restored to them by contact with heated charcoal, a body which was supposed to be specially rich in the hypothetical inflammatory principle.

Whilst we believe that the presence of no substance such as phlogiston is necessary for the production of fire, and that during the manifestation of the phenomena of com-

[1] φλογιστός, anything *set on fire*.   [2] Died at Berlin, 1734.

bustion no loss of material occurs, yet it has been very justly pointed out that bodies, when they burn, do in truth part with something, and that is the potential energy or power of doing work which belongs to a state of chemical isolation. (See Chap. XIX.)

In order that combustion may commence in air the temperature of combustible bodies must in general be raised. The temperatures required in different cases are very diverse. Thus phosphorus, which liquefies at 44°, can scarcely be melted in the air without inflammation. Carbon disulphide vapour mixed with air takes fire if a glass rod heated to about 150° is brought into contact with it. Sulphur begins to burn at about 250°—far below its boiling-point; whilst carbon and many hydrocarbons require a red heat.

The temperature produced when the process of burning is once established is in general higher than that which is requisite for the commencement of combination. This difference is illustrated by the action of platinum upon a mixture of hydrogen or coal gas and air. If a warm slip of clean platinum-foil or a coil of plantinum wire is held in a current of such mixed gases, the temperature of the metal rises rapidly, in consequence of combination taking place between those portions of the gases which are in immediate contact with it, combination extending to the surrounding mass only when the temperature reaches a certain point, and the platinum is nearly white-hot. Similar phenomena may be observed in other cases when the heat evolved in the early stages of the process is allowed to accumulate. The spontaneous ignition of phosphorus, of finely pulverulent iron or lead (pyrophori), and of heaps of oily rags, may be referred to this cause.

The exact temperature of flame is difficult to determine, and is liable to vary. The temperature of a hydrogen flame, burning in air, has been estimated at about 2080° C., but when the flame is fed with pure oxygen its temperature rises to upwards of 4000° C. This is easily explained by the fact that in atmospheric air the oxygen is mixed with four times its bulk of nitrogen, which contributes nothing to the chemical action, and which, being raised to the same temperature as the other gases present, consumes a great deal of heat. A temperature still higher is produced when a mixture of hydrogen and oxygen in due proportions is fired in a closed vessel, so that the heated gases are not allowed to expand. This expansion against atmospheric pressure is work the performance of which involves the consumption of heat. The temperature produced by the explosion of oxygen and hydrogen in a closed vessel has been estimated at about 5250° C.

But although the temperatures producible by the same combustible under various circumstances are different, the actual amount of heat evolved in the combustion of the same weight of a given substance is always the same. This statement can of course be accepted only on condition of uniformity in the circumstances attending the experiment. Thus it will appear from the table given below that, as in the case of carbon, the different allotropic modifications of the same substance may give rise to appreciably different amounts of heat.

I. *Units of Heat developed by Combustion of Equal Weights of Elements in Oxygen.*

Substance burned	Product	Grams of water heated 1° C. by burning 1 gram of each substance
Hydrogen	Water	34,034
Carbon—		
a. Diamond	Dioxide	7,770
b. Natural graphite	,,	7,797
c. Wood charcoal	,,	8,080
Sulphur (native)	,,	2,220
Phosphorus (common)	Pentoxide	5,747
Zinc	Oxide	1,330
Iron	Peroxide	1,582

II. *Units of Heat evolved by Combustion of Atomic Weights.*

Name of element	Weight in grams	Product	Grams of water heated 1°C. by the combustion
Hydrogen	1	Water (liquid)	34,034
,,	1	Hydrochloric acid (gas)	22,000
,,	1	Hydrobromic acid (gas)	8,440
Carbon—			
a. Diamond	12	Dioxide (gas)	93,240
b. Graphite	12	,, ,,	93,560
c. Charcoal	12	,, ,,	96,960
Sulphur (native)	32	Dioxide (gas)	71,042
Phosphorus (common)	31	Pentoxide (solid)	178,157
,,	31	Pentachloride (solid)	107,740
Tin	118	Dioxide (solid)	135,360
,,	118	Tetrachloride (liquid)	122,880

The quantity of heat absolutely evolved also depends partly upon the physical condition of the products of combustion. Thus the number 34,034, which expresses the heat evolved in the combination of one part by weight of hydrogen with eight parts of oxygen, represents not only the heat of chemical action, but the heat (amounting at the temperature of the experiment to about 5,500 units) which is produced by the liquefaction of the resulting nine parts of steam.

This relation of the amount of heat evolved to the physical state of the resulting compounds is further indicated by the results exhibited in the following table:—

Substance burned	Weight burned	Product	Units of heat evolved
Tin	Sn = 118	SnO_2	135,360
Stannous Oxide	SnO = 134	SnO_2	69,584 × 2 = 139,168
Copper	Cu = 63·5	CuO	38,304
Cuprous Oxide	½Cu_2O = 71·5	CuO	18,304 × 2 = 36,608
Graphite	C = 12	CO_2	93,560
Carbonic Oxide	CO = 28	CO_2	67,284 × 2 = 134,568

Here we find that when solid tin or copper is converted into its highest oxide, the amount of heat developed is, practically speaking, twice as great as the amount of heat developed in the conversion of the lower into the higher oxide. In other words, the two successive stages of oxidation, both of which result in the formation of solid products, are marked by the evolution of equal quantities of heat.

The case of carbon is different. In the first stage of oxidation the process involves the conversion of solid carbon into gaseous carbonic oxide, whilst in the second stage the carbonic oxide which is burnt, and the carbonic anhydride which is formed, are both gaseous. There is no change of state. Hence the quantity of heat which is developed in the latter operation is nearly two-thirds, instead of only one-half, the total quantity evolved in the formation of the same weight of carbonic anhydride from solid carbon.

In order, therefore, to calculate the actual amount of heat obtainable by burning a given combustible, it is necessary to take these and other circumstances into consideration. The following examples, which are unencumbered by small corrections, and in which it is assumed that no heat is lost by radiation or conduction, will serve to indicate the general nature of such calculations.

The combustion of 1 part by weight of wood charcoal evolves 8,080 units of heat. That is to say, 1 kilogram of charcoal would heat 8,080 kilos. of water from 0° to 1°, or 1 kilo. of water from 0° to 8080°. 12 kilograms of charcoal produce 44 kilos. of carbonic anhydride, or 1 kilo. produces 3·67 kilos. ; and if the heat produced by the combustion is communicated to this quantity of carbonic anhydride, and not to water, the temperature would be $\frac{8080}{3·67}$, or 2202°, if the specific heat of carbonic anhydride were the same as that of water. But the specific heat of carbonic anhydride is only ·2164, when that of water is 1. Hence the temperature of the carbonic anhydride is $2202 \times \dfrac{1}{.2164}$, or $10175°$, when the carbon is burnt in oxygen.

Now if the combustion is performed in atmospheric air, which contains 77 per cent. of nitrogen, much heat is consumed in raising the temperature of this nitrogen.

The 2·67 parts of oxygen required for the combustion of one part of carbon are accompanied by 8·93 parts of nitrogen, the specific heat of which is ·2438. Therefore, when the combustion of carbon takes place in air, the temperature of the resulting mixture of gases cannot be higher than

$$\dfrac{8080}{(3·67 \times ·2164) + (8·93 \times ·2438)} = 2720°\ C.$$

In practice, the temperature is not so high as this, partly because some heat is lost by radiation, some by conduction through the solid unburnt charcoal, partly because an excess of atmospheric air over and above that actually required mingles with the products of combustion, and partly also because, in all probability, the specific heats are not constant, but increase as the temperature is higher.

CHAPTER XXI.

ISOMERISM.

SEVERAL bodies, though differing more or less in properties, may have the same composition. In such cases they are said to be *isomeric*. The differences observed among isomeric bodies sometimes extend only to their physical characteristics, sometimes to their chemical properties.

Several cases require, therefore, to be considered.

PHYSICAL ISOMERIDES.

Sulphur, when crystallised from carbon disulphide, yields rhombic octahedra, the relative density of which is 2·07;

whereas, if melted and allowed to cool, it crystallises in oblique rhombic prisms, having the relative density 1·98. The prismatic variety soon changes spontaneously into the octahedral, which is the stable form, at the same time evolving heat. Sulphur is said to be *dimorphous*, as it crystallises in two forms, and these two modifications are often spoken of as allotropic states of the element.

Many other examples might be cited of the same substance assuming different crystalline forms, the change of structure being almost invariably attended by differences of density and solubility. One form is generally less stable than the other, and sooner or later, especially under the influence of change of temperature, is converted into the permanent variety.

When, as in these cases, two bodies chemically alike exhibit slight differences of physical characters—such as solubility, crystalline form, relative density, or action upon light—they may be regarded as one and the same substance, though more or less disguised, and such bodies may be distinguished as *physically isomeric*.

As examples of physical isomerism we have the two varieties of native calcium carbonate, arragonite and Iceland spar, and the curious instance of change of colour accompanying change of crystalline structure in the two modifications of mercuric iodide. Common cane sugar, which melts at 160°, is converted at that temperature into an amorphous modification (barley sugar), which slowly recovers its crystalline structure if kept for some months.

The difference in all these cases probably arises not from any difference in chemical composition or constitution, but from the various modes in which the molecules of the bodies are associated together; such modification being connected with some peculiarity in the circumstances attending their formation.

That a change of molecular structure is accompanied by a corresponding change of physical characters is proved by the well-known fact that when a piece of glass is strongly compressed, either by mechanical means or by suddenly cooling it from a high temperature, it acquires a power of polarising light, which it loses when the pressure is removed.

POLYMERIDES.

Bodies containing the same elements united in the same proportion, but having different molecular weights, belong to this class of isomerides. The following are some examples:—

Sulphur boils at 440°, and is converted into an orange-coloured vapour, the density of which, when taken at about 500°, is three times as great as it should be theoretically; whilst at 1000° it is only 32 times as great as that of hydrogen at the same temperature and pressure, thus conforming with the ordinary rule. There seem, therefore, to be two varieties of the gaseous sulphur molecule, one of which is polymeric with the other. They may be represented by the formulæ

$$S_2 \text{ and } S_6.$$

Ozone, the molecule of which has been shown by various facts and arguments to have the formula O_3, or $\frac{3}{2}O_2$, may also be regarded as polymeric with ordinary oxygen, O_2.

The two varieties of nitric peroxide, NO_2 and N_2O_4, have already been described (p. 168). The latter is polymeric with the former.

Of carbon compounds exhibiting similar relations, the hydrocarbons of the C_nH_{2n}, or olefine series, afford a prominent instance. The formulæ of these bodies are all multiples of the first, methylene, CH_2, which, however, is not known in the free state. Being at once polymeric and homologous with one another, they exhibit a regular gradation in their boiling-points and densities, and form similar chemical compounds.

Name	Formula	Boiling-point
Ethene or ethylene	C_2H_4	Gas, liquefiable only under great pressure
Propene or propylene	C_3H_6	$-17\cdot 8$
Tetrene or butylene	C_4H_8	$3°$
Pentene or amylene	C_5H_{10}	35
Hexene or hexylene	C_6H_{12}	68—70
&c.	&c.	&c.

Further illustrations are supplied by the two chlorides of cyanogen, $CNCl$ and $C_3N_3Cl_3$; by cyanic acid, $CNOH$, and cyanuric acid, $C_3N_3O_3H_3$; also by the modifications of aldehyd, C_2H_4O, paraldehyd, $(C_2H_4O)_3$ or $C_6H_{12}O_3$, metaldehyd, $(C_2H_4O)_n$, and many others.

METAMERIDES.

We have now to consider a kind of isomerism which occurs very frequently among carbon compounds. The nature of the phenomenon will be understood by comparing together several such compounds as the following, all of which are represented by the same empirical formula :—

1. Propionic acid, $C_3H_6O_2$, is a crystallisable acid, which, after melting, boils at 140°. It is monobasic, forming one salt only with each of the metals sodium, potassium, and silver. Its rational formula may, therefore, be written thus : $HC_3H_5O_2$.

2. Ethylic formate is a colourless aromatic liquid, which boils at 56°. When heated with caustic potash it is resolved into ethylic alcohol and potassium formate. This mode of decomposition is recorded when we write the formula thus : $C_2H_5.CHO_2$.

3. Methylic acetate is a colourless, volatile liquid, which also boils at 56°, but when decomposed by an alkali it yields wood spirit (methylic alcohol) and an acetate. So its formula must be $CH_3.C_2H_3O_2$.

These three compounds, which have the same composition and molecular weight, but differ in the nature of the products they yield, when decomposed or acted upon by chemical agents, are said to be *metameric* with one another.

Many cases similar to the last two may easily be found amongst ethereal salts (compound ethers), ketones, and other bodies, the molecules of which consist of two compound radicals united together by oxygen or a bivalent group. Pairs of such bodies may be called reciprocal metamerides, because the excess of carbon and hydrogen in one of the radicals is made up for by a corresponding deficiency in the other. General formulæ for such pairs of isomerides among the compound ethers might be written thus:—

$$\left.\begin{array}{c}C_nH_{2n-1}O\\C_pH_{2p+1}\end{array}\right\}O \quad \text{and} \quad \left.\begin{array}{c}C_pH_{2p-1}O\\C_nH_{2n+1}\end{array}\right\}O$$

In many cases an extraordinary resemblance in physical properties may be observed in comparing together two bodies of this kind.

Another instructive example of metameric relations is presented by the two classes of alcoholic cyanides.

1. *Cyanides.*—By distilling ammonium acetate with phosphoric anhydride (a substance which has an extraordinary affinity for water) the ammonium salt is converted into a volatile body, long known as acetonitril, and boiling at 77°.

$$NH_4C_2H_3O_2 - 2OH_2 = NC_2H_3$$

When acted upon by boiling alkali it yields up its nitrogen in the form of ammonia, and re-generates an acetate.

$$NC_2H_3 + OKH + OH_2 = NH_3 + C_2H_3(OK)O$$

This reaction is explained by supposing that in acetonitril or methyl cyanide the two atoms of carbon are in direct union with each other, as represented by this diagram:

$$N \equiv C - CH_3.$$

2. *Isocyanides or Carbamines.*—These compounds under-

go a different transformation when acted upon by hydrating agents. Notwithstanding that they have the same composition as the cyanides, they are scarcely affected by alkalies, though when boiled with diluted acids they are readily converted into formic acid and bases, in which the nitrogen is associated with part of the carbon. For example, methyl isocyanide treated in this manner yields methylamine and formic acid.

$$NC_2H_3 + 2H_2O = N\begin{cases}H \\ CH_3\end{cases} + CHO.OH$$

Methyl isocyanide.   Water.   Methylamine.   Formic acid.

In these compounds, then, the nitrogen probably forms the link between the two atoms of carbon, in the manner shown by the following graphic formula:—

$$C \equiv N - CH_3$$

One very remarkable kind of metamerism is exhibited by those numerous carbon compounds which possess the power of rotating the plane of polarisation of a ray of polarised light, such as the sugars and allied compounds, the varieties of tartaric acid, of amylic alcohol and their derivatives. In all such compounds, the constitution of which is known, it seems to have been established, according to MM. Van t'Hoff and Le Bel, that the molecule contains at least one atom of carbon, having its four units of valency occupied in four different ways. Thus malic acid contains

$$CO_2H - \mathbf{CH}(OH) - CH_2 - CO_2H.$$

In this formula we see one of the carbon atoms (printed in heavy type) combined with CO.OH, with H, with OH, and with CH_2. This atom of carbon is said to be *asymmetric*.

In a great many of these optically active compounds pairs of isomerides are known which, whilst exhibiting almost complete agreement in chemical properties and in most of their physical characters, possess the power of rotating the polarised ray in opposite directions. Thus the dextro- and lævo-rotating tartaric acids have the same constitution, and no difference between them can be exhibited by the usual formula—

$$CO_2H-(CH.OH)-(CH.OH)-CO_2H,$$

which contains two asymmetric carbon atoms. According to the hypothesis of Van t'Hoff the four radicles united to each asymmetric atom are placed at the angles of a regular tetrahedron, the atom of carbon occupying the centre. Two arrangements of the attendant radicles are then possible, the one being a reflection of the other, but not superposible upon it, as shown in the following diagram.

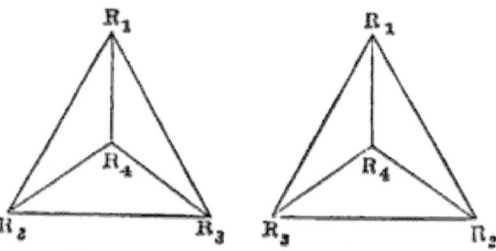

When a compound, the formula of which contains one or more asymmetric carbon atoms, is found to be inactive to the polarised ray, it is usually resolvable into equivalent quantities of two active isomerides, the one rotating to the right, the other to the left. Thus racemic acid is made up of dextro- and lævo-tartaric acids, which may be distinguished thus:

$$C(H.OH.CO_2H)-C(H.OH.CO_2H)$$
$$C(H.CO_2H.OH)-C(H.CO_2H.OH)$$

A fourth acid, called mesotartaric acid, which has long been known, is optically inactive, but not resolvable like racemic acid. It may be represented on the same hypothesis thus:

$$C(H.OH.CO_2H)-C(H.CO_2H.OH),$$

where the effect of one half of the molecule is compensated by the other half.

ALLOTROPY.

Many of the elements are known to exist in the form of two or more modifications, which are very different in physical properties and to some extent also in chemical behaviour. One or two examples have been already referred to under the head of physical isomerism and of polymerism. It is probable that a closer examination of the different cases of allotropy known among the elements would enable us to range them all in one or other of these classes. At present, however, our knowledge will not allow us to adopt with any degree of confidence a final decision upon this point.

In the earlier part of this chapter prismatic sulphur is described as a physical or mechanical modification of the octahedral form of the element, and this is probably correct. But these two are not the only varieties of which this body is susceptible. By heating melted sulphur to a temperature of 240° to 250° it becomes extremely viscid, and if cooled suddenly whilst in this condition the viscid consistency is retained, and the product is a tough elastic solid quite different in aspect from ordinary sulphur. In this state it is insoluble in carbon bisulphide, a liquid which takes up octahedral sulphur very freely.

After keeping a few hours it becomes brittle and crystalline, and recovers its solubility in the usual solvents. The same change may be brought about in a few minutes by plunging it into hot water.

In these transformations, and in the insoluble character of this plastic sulphur, we may trace a close resemblance to the modifications to which certain carbon compounds, such as aldehyd, cyanic acid, and other bodies, are subject, and which are known to be the effect of polymeric changes. It seems not unreasonable to consider that the production of plastic sulphur may be brought about in the same way.

Phosphorus presents us with an example of a somewhat

similar kind. This element in its ordinary state is at common temperatures a solid of waxy consistence, which becomes brittle at low temperatures. Its relative density is 1·82. It melts easily, dissolves in carbon disulphide, and by sublimation or solution it may be obtained in brilliant crystals in the form of regular octahedrons and dodecahedrons. When this body is heated to a temperature approaching 250° it is slowly transformed into a dull red powder or mass, of relative density 2·14, which is no longer soluble in carbon disulphide. It shows itself also in many ways less inclined to enter into chemical combination than common phosphorus, being far less easily inflammable and oxidisable, and unaffected by hot alkaline solutions.

The explanation of this appears, at least very probably, to be that the molecules of the ordinary phosphorus combine together into more complex groupings to form the allotropic molecules, and so expend part of their chemical energy.

Carbon is another element which assumes several distinct forms, the relations of which are of great interest. We may divide these various modifications into two distinct groups, the crystalline and amorphous.

Crystalline carbon is dimorphous. In one form it constitutes the diamond, which crystallises in octahedral forms of the regular system, and has a relative density on the average of 3·3. The other is graphite, or, as it is frequently called, plumbago or black-lead, the crystalline form of which, hexagonal plates, is quite incompatible with the form of the diamond. The average density of graphite is 2·2. If now we review the results which have been obtained by burning the different forms of crystalline carbon, we find that, allowing for slight experimental errors, the amount of heat evolved by the combustion of equal weights of diamond and graphite are practically the same. Twelve grams of each substance burnt in excess of oxygen disengage enough heat to raise the temperature of about 93,300 grams of water one degree, or, as it is usually expressed,

93,300 units of heat are evolved. The exact numbers in each case are as follows :—

Diamond	93,240
Natural graphite	93,560 ⎫ mean
Graphite from iron	93,140 ⎭ 93,350.

The smallness of the difference observed would lead one to the belief that graphite and diamond possess the same atomic structure, and that they owe their peculiarities to different arrangement of their molecules—that, in short, they belong to the class of physical isomerides, were it not for some remarkable facts in connection with their behaviour under the influence of chemical reagents. A mixture of nitric acid and potassium chlorate has no action on the diamond, even in the state of the finest dust, but under the influence of this powerful oxidising mixture graphite is converted into a yellow crystalline substance, called by Sir B. Brodie, who discovered it, graphic acid. This compound contains $C_{11}H_4O_5$ (Brodie), and when heated it decomposes violently, leaving a black graphitic residue, which still retains oxygen and hydrogen.

Amorphous carbon may be obtained by a great variety of processes, and in each case the product exhibits more or less distinctly marked peculiarities.

But neither vegetable nor animal charcoal, lamp-black, coke, nor gas carbon yields by the action of potassic chlorate any substance of the nature of graphic acid, but only black soluble substances of indefinite composition. Twelve grams of wood charcoal give out 96,960 units of heat when burnt so as to form carbonic anhydride, and other kinds of charcoal, when deprived as completely as possible of hydrogen and oxygen, give numbers closely agreeing with this. Taking the average heat of combustion of crystalline carbon as 93,300 units, it is obvious that there is too great a difference here to be fairly accounted for by the hypothesis of experimental error, and consequently that there is some essential difference in the constitution of crystalline and amorphous carbon. The question whether this difference is sufficient to indicate a polymeric relation between these bodies remains to be answered. Silicon and boron form allotropes, which are analogous to those of carbon, and concerning which the same questions may be propounded.

The study of a great number of cases has led to the discovery that the formation of any two isomeric bodies always involves the consumption of different amounts of heat. Also, that when these bodies are burnt, or otherwise similarly decomposed, the disruption of their molecules is attended by the evolution of different amounts of heat.

This is nearly equivalent to saying that in order to produce equal weights of two isomerides different amounts of work must be expended in the two processes, and that different amounts of energy are stored up in the products.

How is this energy disposed of? According to one view, and adopting the molecular theory, we may reply that the energy is employed in communicating to some atom or atoms within the molecule a new kind of motion whereby it acquires new chemical functions, and this change we figure to our minds, and render intelligible by the hypothesis that in the transformation of a body into its isomeride the position of certain atoms contained within the molecule is changed. We endeavour to represent this change of function by altering the arrangement of the symbols which go to make up the formula of the body. Examples of this will be found freely scattered through these pages; but as an additional illustration we might refer to an interesting case of isomeric change observed by Hofmann.

It was found that methyl-aniline,

$$(C_6H_5)'-N{\Large\langle}{{CH_3}\atop{H}}$$

by protracted heating to a high temperature, is converted into toluidine,

$$(CH_3)\!-\!(C_6H_4)''\!-\!N\!\!<^H_H$$

In this metamorphosis an atom of hydrogen and an atom of methyl, CH_3, appear to exchange functions, and in order to record this exchange the formulæ are written in the above or some similar manner, though it by no means follows that we are to infer an exchange of *place*.

But in expending energy upon some part of a molecule it is not very probable that the energy of the molecule as a whole remains unaltered. Still, it is conceivable that in some cases it may be so, and a careful comparison of the chemical properties of a great many pairs of isomeric bodies would be of the highest interest, as in this way we might arrive at a solution of the question whether bodies possessing a great store of energy are really more active in their chemical behaviour than others in the production of which a smaller amount of energy has been consumed.

CHAPTER XXII.

THEORIES REGARDING THE NATURE OF CHEMICAL AFFINITY.

THE belief that chemical compounds are formed by the union, not of masses, but of atoms of matter, is generally adopted by chemists. But why they unite, why a given atom possesses the power of selecting among other atoms presented to it, or why it habitually associates with one, two, three, or more atoms of another kind (valency), are questions which in the present state of knowledge cannot be answered save by hypothesis.

Before entering upon the discussion of various theories which have been at different times proposed to account for the phenomena of chemical combination and decomposition, it will perhaps be well to recapitulate (see also Chap. XVII.) the chief characteristics which are attributed to chemical affinity.[1]

1. It acts only at inappreciable distances, smaller than any which can at present be measured. Atoms chemically united are probably not in actual contact with each other, but are separated by a distance which is at least several times their own diameters. If this is true, the attraction which they exert upon each other must be believed to involve some action in the medium which fills the space between them, as in the case of other attractions at a distance, such as that due to electrical or magnetic induction.

2. It is selective.

3. Chemical change, whether combination or decomposition, is attended by a concentration or dissipation of energy, indicated usually by absorption or evolution of heat.

4. Chemical affinity differs from gravitation not only in not operating at all distances, but in the fact that two or more atoms, having combined together, lose more or less completely the power of attracting other atoms. Gravitation acts not only between a given body and one other, but between that body and all others, so far as we know, in the universe.

5. Chemical compounds are broken up by heat; also (6), according to common belief, by electricity.

Three chief hypotheses have been suggested. The first assumes that chemical combination is the result of an 'attraction' of a peculiar kind which operates only at very short distances. As already stated, this implies a condition of strain in the 'ether' or medium intervening, and until the relation of the intensity of the attraction to the distance between

---

[1] The word affinity seems to have arisen from a notion current among the early chemists, that when two bodies are capable of combining, there must be some resemblance or affinity between them, or that they contain some principle common to them both.

any two atoms is known there is nothing to distinguish this hypothesis from the idea of universal gravitation.

The electro-chemical theory of chemical combination has attracted much greater attention. Immediately after the invention of the voltaic pile it was discovered that the passage of a current of electricity through water and various metallic salts, either dissolved in water or melted by heat, decomposes them, their constituents appearing at the opposite electrodes; that hydrogen and the metals are always liberated at the negative, whilst oxygen, chlorine, and acidulous bodies generally are found at the positive pole. Later, Faraday established two chief quantitative laws, from which we know that the amount of decomposition in a given case is directly proportional to the quantity of electricity flowing through the circuit, and that if the same current passes through a succession of decomposable substances or 'electrolytes,' the amount of each element liberated is proportional to its chemical equivalent (Chap. XIX., p. 184). To separate a univalent atom, such as Cl, from a compound requires a certain fundamental unit of electricity, the value of which has been approximately determined. To separate a bivalent atom, such as oxygen, requires twice as much, whilst a trivalent atom requires three times as much, and so on, multiples of the unit charge being always employed, but never fractions of it. Hence it has been suggested that electricity, like matter, is atomic. A hypothesis relating to the mechanism of electrolysis was devised by Grotthus in 1805, and since that time has been very generally accepted. In a liquid, such as a solution of hydrochloric acid, each molecule of the electrolyte is supposed to be made up of two parts—one having a charge of positive, the other of negative electricity, in virtue of which these constituents are united together. According to this hypothesis, then, chemical affinity is nothing but the attraction of oppositely electrified particles. When a current is sent through such a liquid, the first effect is the establishment of a great number of chains of molecules in which the positive elements all face in one direction, the negative elements in the other. Then the metal plates at different potentials which form the electrodes attract the atoms of opposite electric name from the two ends of the polar chains, and detach them from the state of chemical combination in which they previously existed.

The following diagrams serve to illustrate the successive conditions supposed to prevail in the liquid. In the first the molecules of the electrolyte are moving about in the liquid mass in every direction. Then these molecules

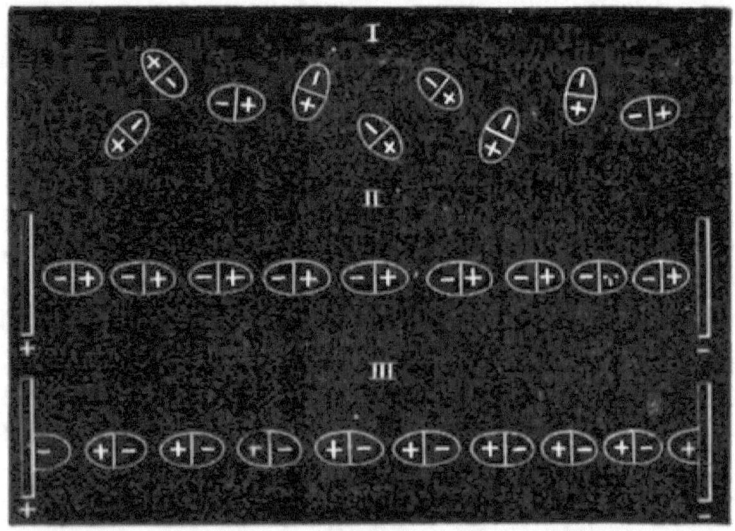

arrange themselves into chains in the order shown in II, and in III the disruption of the end molecules occurs, and is followed by a change of partners throughout the length of the chain. Thus the disengagement of hydrogen at the negative electrode, and of an equivalent quantity of chlorine at the positive, without any apparent change in the intervening mass of liquid is accounted for. There are, however, many difficulties attending this hypothesis which have not yet been explained away. Some of these have already been referred to in Chap. XIX. Thus it has been shown that pure anhydrous liquid hydrochloric acid is practically an insulator or di-electric, and is not decomposed by the most powerful current yet applied. Pure water also resists the current so powerfully, that it is even now doubtful whether water entirely free from other substances is capable of electrolytic decomposition. Nevertheless, when hydrochloric acid and water are mixed together, a liquid is obtained which is decomposable with the greatest ease. Again, the hypothesis assumes that the constituents of an electrolyte are united in consequence of charges of electricity resident in the atoms. If these charges are supposed to be inseparable from the atoms, the latter, when liberated from a compound, should cling persistently to the surface of the electrode. This, however, is not the case, for such substances as hydrogen, oxygen, or chlorine either dissolve in the liquid or escape in bubbles. On the other hand, if the charges of the several atoms are given up to the surface of the electrode at which the atoms are set free, the coupling of the atoms into chemically neutral molecules of hydrogen, H_2, oxygen, O_2, or ozone, O_3, &c., cannot be referred to charges of electricity. Moreover, since all atoms of any one element, say hydrogen, liberated by electrolysis, must be charged equally with the same kind of electricity (otherwise they would not all pass to the same electrode), there is no reason to expect that pairs of them would cling together, but rather that they would repel each other, as similarly electrified masses commonly do. Another objection to such an electrical theory of combination, as long ago pointed out by Dumas, arises out of the fact that the halogens, strongly electro-negative as they usually are, can replace hydrogen, atom for atom, in carbon compounds, and the physical and chemical characters of the substitution products closely resemble those of the substances from which they are formed. (See, however, p. 220.)

The third hypothesis relating to the nature of chemical changes is based upon the idea of *atomic motion*, and upon this is founded the view of electrolysis which has been proposed by Clausius. As already explained (in Chap. I. and elsewhere), there are good reasons for believing that the molecules of liquids and gases subsist unceasingly in a state of motion. Their agitation is increased by heat, diminished by cold. In the course of this dance in which they are engaged, and the numerous encounters which must occur amongst them, it is conceivable that some of the molecules get broken up into atoms or atomic groups, which for a while wander about until they encounter some other atom or atomic group with which they can unite. If the original body was homogeneous, the molecules which are thus reproduced are of the same kind as the original molecules. So long as this work of reproduction goes on at the same rate as the destruction—that is, so long as, in a given interval, the number of molecules decomposed and the number of molecules recomposed is the same—no change occurs in the properties of the body, because the average composition of the mass remains the same. In a mass of hydrochloric acid gas, for example, it is conceived that if it were possible to submit it to such a scrutiny, the greater part of the mass taken at any instant would be found to consist of molecules, each made up of an atom of hydrogen and an atom of chlorine; but that with these there would be associated a certain number of free atoms of hydrogen and chlorine interspersed amongst them. In the next instant many of these free atoms would be seen yoked together again, whilst their places would be supplied by the disruption of fresh molecules.

A similar condition is supposed to prevail in a liquid, so that the separation of the atoms of an electrolyte is supposed to be an effect of collisions among the molecules of the compound, whilst the action of the current is merely to direct the already separated atoms to the opposite electrodes. Thus an atom of hydrogen being within a certain distance of the positive electrode may be supposed to receive a charge of positive electricity, the possession of which causes it to be attracted to the surface of the negative electrode, where it arrives in company with other atoms similarly charged. On reaching the electrode the atoms give up their charges to it, and then pair off to form molecules.

Conformable with this hypothesis we have the facts already mentioned (pp. 184-5), that many pure substances are not electrolytically decomposed until they are dissolved in a liquid such as water; and that whereas metallic conduction is diminished by rise of temperature, electrolysis is usually promoted by raising the temperature of the liquid. Both these conditions may be assumed to favour dissociation. If electrolysis is merely a kind of electrical convection in which previously dissociated atoms alone are concerned, the electrical hypothesis of chemical affinity falls to the ground.

Whilst for the more stable forms of chemical combination, such as give rise to the hosts of oxides, chlorides, salts, and carbon compounds, the law of definite proportions forms a safe criterion by which the operation of chemical affinity may be recognised, many facts have been discovered, especially of late years, which tend to support the view that chemical combination is not always distinguishable from mere adhesion; in fact, that these two forms of union merely represent the extreme cases of a long series of phenomena which pass by insensible degrees into one another.

A short statement of some of these facts may be appropriately mentioned in this place.

Chemical affinity is elective. So is adhesion. Thus glass is wetted by water but not by mercury. Again, if a drop of a solution containing a mixture of cadmium and mercuric chlorides is placed upon filter-paper, and allowed to spread, it will be found that the cadmium extends further into the paper than the mercury, the latter remaining in the centre of the blot whilst the former spreads to the edge.

P

Chemical combination is often supposed to be indicated by rise of temperature. But when fine insoluble powders, such as silica, are wetted with water, alcohol, or benzene, a rise of temperature is observed, although no definite compound appears to be formed.

It might be imagined that the relatively greater stability of chemical union would be a sufficient guide, if such facts as the copious absorption of hydrogen by palladium, and the condensation of bromine vapour, ammonia, and other gases by porous bodies, and especially by hard charcoal, did not show that so-called adhesion may be equally persistent. The film of air which covers the surface of glass cannot at common temperatures be removed even by the mercury pump, which produces a very perfect vacuum, unless at the same time the glass is heated to near redness. The tenacity of the combination in this case greatly exceeds that of many recognised cases of chemical combination. Relative stability evidently fails as a test when we consider the properties of such a series of compounds as

SCl_4, PH_4Br, PCl_5, NH_4HS, NH_4Cl, Hg_2Cl_2, $CaCO_3$, SiO_2,

in which we have at one end compounds, the constituents of which cannot remain united even at usual atmospheric temperature and pressure, whilst at the other end there is very great stability.

Again, chemical compounds may be decomposed not only by the operations usually considered to be exclusively chemical, but also by agents commonly regarded as non-chemical or mechanical. Thus lead acetate and lead nitrate, tartar emetic, or ammonio-sulphate of copper, may be completely deprived of their respective metals if a solution of each salt is shaken up with pure carbon from bone charcoal.

Many porous substances, such as paper or plaster, are also capable of removing water of crystallisation from such compounds as crystallised cobalt chloride.

Beginning, then, with ordinary cohesion, in virtue of which similar particles are attached to one another, and adhesion—which is the name given to the unknown cause of the union of dissimilar particles—and passing in review the various phenomena of absorption, occlusion of gases by solids, and solution, we arrive by imperceptible degrees at the compounds which have been long distinguished by many chemists as 'molecular' compounds, in which entire molecules have been supposed to be united together by a special kind of chemical affinity. This distinction is, however, now rapidly disappearing, as it is acknowledged that there is no character by which compounds of water or of ammonia with salts, of iodine with iodides, or of salts with one another, can be sharply distinguished from those in which atoms are supposed to be linked together by ordinary chemical affinity.

The more stable forms of combination seem to result from a comparatively close approximation of the constituent atoms or molecules, which is indicated by differences of density. Thus the specific volume of 'water of halhydration' (as it was called by Graham) in the sulphates of the magnesium group is less than the 'water of crystallisation,' each additional molecule of which occupies a gradually increasing volume.

The following table shows the mean specific volumes of $MSO_4 nH_2O$, where M is Cu, Mg, Zn, Ni, Co, Mn, or Fe.

Value of n	Mean Specific Volume of $MSO_4 nH_2O$	Molecular Volume of nth mol. of H_2O
0	44·8	
1	55·5	10·7
2	68·8	13·3
3	83·3	14·5
4	98·7	15·4
5	112·9	14·2
6	130·0	17·1
7	146·1	16·1

These last figures approach the specific volume of ice, which is $\frac{\text{Mol. Wt.}}{\text{Dens.}} = \frac{18}{\cdot 92} = 19\cdot 6$, that of liquid water being 18.

When these salts are dissolved in water the water of crystallisation acquires the same volume per molecule as the rest of the water in the solution.

Again, when lead replaces silver, or potassium replaces sodium in the nitrate, or when chlorine replaces bromine or iodine in combination with another element, contraction occurs. And in general contraction is observed when an element of reputed strong affinity (as indicated by the amount of heat evolved when it enters into combination) takes the place of one of reputed smaller affinity. The following examples will suffice :

Molecular Weight	Density	$\frac{\text{Mol. Wt.}}{\text{Dens.}}$ = Mol. Vol.
$AgCl = 143\cdot 5$	$5\cdot 517$	$26\cdot 0$
$\frac{1}{2}PbCl_2 = 139$	$5\cdot 78$	$24\cdot 0$
$NaI = 150$	$3\cdot 45$	$43\cdot 5$
$NaBr = 103$	$2\cdot 952$	$34\cdot 8$
$NaCl = 58\cdot 5$	$2\cdot 148$	$27\cdot 2$

As to the relative strength or force of affinity between two atoms in different cases, there is much apparent conflict of evidence. Common experience seems to indicate that the strength of the affinity between different atoms varies according to their nature. Thus an iodide is decomposed by bromine or by chlorine with liberation of iodine. Here, however, the student must be again reminded that the change is not correctly represented as

$$HI + Br = HBr + I.$$

And though at first sight it seems to depend upon the tendency for H to combine with Br in preference to I, to this must certainly be added the affinity of I for I, the interaction probably proceeding in two stages, thus :

$$HI + BrBr = HBr + IBr$$
$$HI + IBr = HBr + II.$$

Little can be inferred even from the replacement of one metal by another, as of mercury by copper, copper by iron, &c., in solutions of their salts, for in all these cases the process is complicated by electrical effects, or by a difference of physical condition between the materials and the products of their interaction, also by the relative masses of the acting ingredients.

It has been supposed that the amount of heat evolved during the formation of series of compounds in which only one element varies might serve as a measure of the relative affinities of the several interchangeable elements, as, for example, in the interchange of metals or of halogens for one another (pp. 180-181). But though unquestionably differences are manifest in such cases it does not follow that the different amounts of energy degraded in the process of combination indicate that the atoms combining come together with different amounts of force. It has been conjectured that the loss of energy may be partly due to changes in the configuration or motion of the atoms themselves. That the amount of heat liberated in a chemical process is no satisfactory measure of the amount of chemical affinity at work is shown by the results of the researches of Julius Thomsen and others on the relative 'avidity' of acids and bases for each other in solution. From these results it is supposed that although acids on neutralisation by an alkali evolve heat in different amounts, as shown in the table on p. 178, the order in which they stand in respect to heat evolved and affinity for the base is not the same. Thus Thomsen concludes that when chemically equivalent quantities of soda, hydrochloric acid (or nitric acid), and sulphuric acid in aqueous solution react upon one another the base is not divided between the acids equally, but in the ratio of about 2 : 1. The activity of hydrochloric acid has also been compared with that of sulphuric acid in causing hydrolysis of milk sugar, and with a similar result. So that although 2HCl is equivalent in neutralising power to H_2SO_4, the dynamical equivalent of H_2SO_4 is HCl. The rate at which various other chemical reactions proceed, such as the etherification of alcohols by acids and the transformation of amides into ammonium salts, has also been determined, and the results seem to point to the conclusion that each substance acts partly in virtue of its mass—that is, the quantity present in the mixture—and according to a specific rate dependent upon its 'affinity.'

There is a further reason why the heat evolved or absorbed in a chemical process is not a strict measure of the energy of combination, for even in those cases in which there is no alteration of physical state, as from solid to liquid, there is *always* a change of volume. In the present state of knowledge no distinction can be made between the chemical and mechanical parts of each process.

On the whole, the hypothesis that chemical combination, as well as adhesion and cohesion, is due to a state of motion of the atoms or molecules engaged in a given operation deserves most consideration. This view conforms with all that is known concerning the mechanical properties of gases; it explains the phenomena of dissociation, and without it the influence of 'mass' upon chemical processes would be unintelligible. The relatively great stability of the molecules of the so-called 'elements' (*e.g.* N_2, O_2, &c.) in which the compound atoms are alike seems also explicable on the hypothesis that this stability is due to a similarity or harmony of their motions rather than to anything in the nature of an electrical charge.

EXERCISES ON SECTION IV.

1. The density of the vapour of ammonium chloride is said to be abnormal. Explain this statement, and describe the experimental evidence upon which it is based.

2. Give some examples of the influence of 'mass' in determining chemical reactions, and explain them.

3. Ammonium carbonate and calcium chloride are dissolved in separate portions of water, and the solutions are then mixed. Express the result by an equation.

Ammonium chloride and chalk, both dry, are mixed together and heated. Express the result by an equation.

How do you explain such apparently inconsistent phenomena?

4. Ferric oxide is heated in hydrogen. Iron is heated in steam. What effects are produced in each case, and how can you explain such apparently contradictory results?

5. Referring to the table of heat of formation of chlorides (pp. 180-181), determine which of the metals, zinc, cadmium, mercury, copper, gold, will decompose aqueous hydrochloric acid, and range them in their apparent order of affinity for chlorine.

6. What substances and what relative quantities of each will be formed by passing the same current through solutions in water of potassium iodide, copper sulphate, hydrochloric acid (strong), and hydrochloric acid (dilute)?

7. A current of electricity is passed simultaneously through solutions of cupric and cuprous chloride. How much copper and how much chlorine are liberated from the cuprous chloride for every molecule of cupric chloride decomposed by the current? In what relation do these quantities stand to the quantity of zinc consumed in each cell of the battery, secondary actions being neglected?

8. Define in a few words the terms *allotropy, metamerism, polymerism*.

9. What weight of water would be heated from $0°$ to $1°$ by the combustion of 1 gram of charcoal in oxygen?

10. What weight of water would be heated from $0°$ to $15°$ by the combustion of 1 gram of hydrogen in chlorine?

11. Calculate the temperature of combustion of phosphorus burning in air.

12. Heat of combustion of hydrogen . 34034 units.
    Latent heat of steam at $100°$ . . . 537 ,,
    Specific heat of steam . . . . ·475 unit.
    Specific heat of nitrogen . . . ·2438 ,,

Composition of air, N 77, O 23 parts by weight.

With these data find the temperature of the hydrogen flame burning in air.

13. What weight of water at 100° would be turned into steam by the combustion of 1,000 kilograms of coal containing—

Carbon	90·7 per cent.
Hydrogen	4·3 ,,
Oxygen and nitrogen	4·0 ,,
Incombustible ash	1·0 ,, ?

Assume the heat of combustion of C and H as given in the table on p. 189. Latent heat of steam at 100° = 537.

14. The heat of combustion of 1 part by weight of marsh-gas is 13,000. What weight of ice will be melted by the heat produced by the combustion of a litre of the gas measured at normal temperature and pressure?

Latent heat of water at 0° = 79.

15. What volume (calculated at 0° and 760 mm.) of producer gas consisting of—

Carbonic oxide	25 per cent.
Hydrogen	5 ,,
Marsh-gas	4 ,,
Carbon dioxide	6 ,,
Nitrogen	60 ,,

must be burned in order to convert 2,000 kilograms of water at 0° into steam at 100°?

Heat of combustion of marsh-gas = 13,000.

SECTION V.

CHAPTER XXIII.

CLASSIFICATION OF ELEMENTS.

THE following list includes all the known elements except some twelve rare metals, the consideration of which, obscure as their characters are, is unnecessary in a work like this. The numbers in the table represent the relative densities of those which are solid, usually in the densest form in which they are known.

I. *Non-metallic or Oxygenic*[1] *Elements.*

Gaseous :
O, N, F, Cl (liquid 1·33).

Liquid :
Br 2·96

Solid :
S (octahedral)   2·07
P (red)   .   2·2
Si (graphitic)   2·49
B (adamantine)   2·68
C (adamantine)   3·5
Se (crystalline) 4·5
I   .   4·95

[1] Oxygenic = acid-producing.

II. *Metalloids or Imperfect Metals.*
Gaseous:
   H

Solid:
   Zr . . . 4·15
   V . . . 5·5
   As . . . 5·9
   Te . . . 6·25
   Sb . . . 6·8
   Sn . . . 7·3
   Mo . . . 8·6
   Bi . . . 9·8
   W . . . 17·6
   U . . . 18·4
   Ti, Nb, Ta (?)

III. *Metals or Basigenic*[1] *Elements.*
   Li . . . ·578
   K . . . ·865
   Na . . . ·97
   Rb . . . 1·52
   Cs . . . 1·88
   Be . . . 1·7
   Mg . . . 1·7
   Ca . . . 1·8
   Sr . . . 2·5
   Al . . . 2·6
   Ba . . . 4·0
   Ga . . . 5·9
   Ge . . . 5·5
   Cr . . . 6·8
   Zn . . . 7·0
   In . . . 7·4

[1] Basigenic = base-producing.

Classification of the Elements.

Fe	7·8
Mn	8·0
Ni	8·6
Cd	8·7
Cu	8·9
Co	8·9
Ag	10·53
Ro	11·0
Pb	11·36
Ru	11·4
Pd	11·8
Tl	11·9
Hg (liquid at 0°)	13·596
Au	19·34
Os	21·4
Ir	21·15
Pt	21·5

This division into three groups is adopted here for purposes of convenience, but the student must not infer that it is absolutely necessary. Indeed, as he goes on he will find that, as in all attempts to classify the things of nature, it is impossible to define precisely a border-line separating a given class of bodies from all others.

DIVISION I.—NON-METALS.

These elements, as a class, are characterised by no generality of physical properties. At common temperatures and pressures four are gases—fluorine, chlorine, oxygen, nitrogen; one, bromine, is a liquid; the rest are brittle solids. Of these, iodine, sulphur, selenion, and phosphorus are fusible and vaporisable; the remaining three are distinguished by infusibility (?), by non-volatility, by abnormal specific heats (Chap. XIV., p. 115), and by affording, in the cases of graphitic carbon and silicon, the only notable examples among the non-metals of electric conductivity.

In their chemical characteristics, however, there is tolerable uniformity. They all combine with hydrogen, forming gaseous or very volatile hydrides; they also combine with metals, often in several proportions; their oxides are either neutral and indifferent bodies, like carbonic oxide, or, the great majority, anhydrides, which by uniting with water form acids.

NON-METALS.—CLASS I : THE HALOGENS.

$$\left.\begin{array}{llll}\text{Fluorine} & . & F & = 19 \\ \text{Chlorine} & . & Cl & = 35\cdot5 \\ \text{Bromine} & . & Br & = 80 \\ \text{Iodine} & . & I & = 127\end{array}\right\} \frac{Cl + I}{2} = 81\cdot25$$

These elements are characterised by a remarkable family resemblance. The last three especially are constantly associated together in nature in the haloid [1] salts of potassium, sodium, &c., and in the ores of mercury, silver, and other heavy metals. They also agree very closely with one another in their general physical characters and chemical deportment.

At ordinary temperatures chlorine is gaseous, bromine liquid, iodine solid; but bromine and iodine are volatile, and yield heavy, coloured vapours, which, when largely diluted with air, have nearly the same odour as chlorine. Each forms with hydrogen a strongly acid compound, which under ordinary conditions is a colourless, fuming, very soluble gas, consisting of equal volumes of hydrogen and the vapour of the halogen, united without contraction. The chlorides, bromides, and iodides of the alkali metals crystallise in the same form, and the isomorphous replacement of the one halogen by another is observed in a great many other cases.

The following table exhibits the formulæ of all the known oxides and acids of chlorine, bromine, and iodine, from which it will be seen that although there are many gaps to

---

[1] ἅλς = sea-salt = common salt.

be filled up, perhaps by future research, the correspondence, so far as it goes, is almost complete :—

Halogen Oxides and Corresponding Acids.

	Cl_2O	Cl_2O_3 (?)	ClO_2		
HCl	HClO	$HClO_2$	$HClO_2$ \} $HClO_3$ \}	$HClO_3$	$HClO_4$

No oxide of bromine known.

HBr	HBrO			$HBrO_3$	
				I_2O_5	
HI	HIO (?)			HIO_3 (?)	HIO_4 (?)
				or $H_2I_2O_6$	or H_5IO_6

The differences exhibited by chlorine, bromine, and iodine are strictly gradational; chlorine, with the smallest atomic weight, being most active, bromine next, and iodine the least energetic of the three. These differences are manifested by their relative activity towards the metals and hydrogen, chlorine displacing bromine, and bromine displacing iodine from such combinations.

Indications of the same differences are afforded by the superior activity of chlorine as a bleaching agent, and by the energy with which it replaces hydrogen in carbon compounds. As in several other cases of nearly allied elements, to be referred to hereafter, the chemical activity diminishes in proportion to the increase of the atomic weight, and rise of boiling-point and density.

The replacement of one or more atoms of hydrogen in a hydrocarbon by an equivalent quantity of one of the halogens produces a neutral substitution compound; but if a similar replacement is effected in the molecule of a body which contains oxygen, the product not unfrequently presents well-marked acid properties. This is the case, for example, with some of the derivatives of phenol.

This oxygenic tendency of the halogens is also indicated by the destruction of basic character in the amines or compound ammonias by the substitution of chlorine, bromine, or iodine for their hydrogen, as is well shown by the chlorinated derivatives of aniline.

Aniline, C_6H_7N, a powerful base.
Chloraniline, C_6H_6ClN, less basic than aniline.
Dichloraniline, $C_6H_5Cl_2N$, feeble base.
Trichloraniline, $C_6H_4Cl_3N$, neutral.

Iodine.—Iodine presents one or two peculiarities which deserve special notice, as they serve to remove it to some slight extent from immediate association with the kindred elements, bromine and chlorine. In the first place, its affinity for hydrogen is decidedly less energetic than that of either of the other two elements. This is indicated first by the fact that iodine does not usually bleach vegetable colours; secondly, that, acting alone, it is incapable of producing substitution derivatives from carbon compounds.[1] Whenever substitution of chlorine, bromine, or iodine occurs, the hydrogen which is necessarily eliminated goes to form the corresponding hydracid. Now, in the case of iodo-substitution compounds, it has been shown that they are all decomposed by the action of hydriodic acid, with reproduction of the original body and free iodine. Hence iodo-substitution compounds cannot be formed by the action of iodine, unless precautions are taken to remove or to destroy the hydriodic acid that may be produced. This is effected in various ways, usually by the action of mercuric oxide or iodic acid. The difficulty may also be got over in some instances by substituting iodine monochloride for iodine. Thus, orcinol acted upon by a solution of iodine chloride gives tri-iodorcinol and hydrochloric acid—

$$C_7H_8O_2 + 3ICl = C_7H_5I_3O_2 + 3HCl.$$

Another distinguishing characteristic of iodine is the intense colour exhibited by the vapour of the element itself, by

[1] See also Chap. XIX., p. 181, Heat of Combination of Iodine with Hydrogen.

its solutions in certain liquids, notably in carbon disulphide, by its compound with starch, and by many iodides, the corresponding chlorides or bromides being either colourless or very pale.

Again, chlorine and bromine are more soluble in water than iodine, and are even capable of forming at low temperatures crystalline hydrates, having the formulæ $Cl_2.10H_2O$ and $Br_2.10H_2O$, no such compound being formed by iodine. On the other hand, the solubility of chlorine and bromine is not increased by the addition of a chloride or bromide to the water in which they are to be dissolved. The solubility of chlorine in aqueous solutions of chlorides is usually less than its solubility in pure water. Iodine, however, is freely soluble in iodide of potassium, and, indeed, produces in this way a black liquid which contains an unstable tri-iodide of potassium, KI_3. This compound forms dark blue deliquescent prisms. The corresponding ammonium salt, NH_4I_3, resembles it closely, and analogous periodides are formed by the organic ammonium bases, some of which form crystals of great beauty, containing two, four, six, or even eight atoms of iodine in addition to the elements of the normal iodide; as, for example, the following compounds derived from tetramethylammonium iodide :—

$N(CH_3)_4I$, $N(CH_3)_4I_3$, $N(CH_3)_4I_5$, $N(CH_3)_4I_9$

The iodates exhibit some anomalies for which there is no parallel among the chlorates. Thus, in addition to the normal potassic iodate, KIO_3, there are two other well-crystallised salts, containing an excess of anhydride, for which it is difficult to find analogues, except, perhaps, among the chromates.

Iodates	Chromates
KIO_3 or $K_2I_2O_6$	K_2CrO_4
$K_2I_2O_6.I_2O_5$	$K_2CrO_4.CrO_3$
$K_2I_2O_6.2I_2O_5$	$K_2CrO_4.2CrO_3$

This tendency of iodine to accumulate in its compounds is just one of those characters which belong especially to multivalent elements, among which iodine seems, on the whole, entitled to be placed.

Fluorine.—This element has been at last isolated (1886) by the electrolysis of a mixture of hydrogen and potassium fluorides. It is described as a colourless gas which decomposes water, forming hydrofluoric acid and ozone. Silicon and boron burn in it, forming fluorides. Fluorine is separated from the other halogens, partly in consequence of the vapour density of hydrofluoric acid, corresponding to the formula H_2F_2 at temperatures much above its boiling-point, partly by its very extraordinary attraction for silicon, partly by the non-existence of any oxide or oxyacid of fluorine. In spite, also, of their general resemblance to the chlorides, bromides, and iodides, individual fluorides differ in many cases from the corresponding chlorides. Thus, fluoride of calcium is insoluble, chloride of calcium very soluble and deliquescent; fluoride of silver is soluble in water, chloride of silver totally insoluble; fluoride of potassium soluble in water, but, unlike the neutral stable chloride, it yields an alkaline solution which probably contains caustic potash and the double hydrogen and potassium fluoride.

$$2KF + OH_2 = KOH + KF.HF.$$

The tendency to produce double salts of this kind has, indeed, led to the idea that fluorine may be really a bivalent element, like oxygen, with the atomic weight 38. The formulæ of the double fluorides would then be comparable with those of oxygen compounds.

Fluorides		Oxides
H_2F	analogous to	H_2O
KHF	,,	KHO
KBF_2	,,	KBO_2
K_2SiF_3	,,	K_2SiO_3
K_2SnF_3	,,	K_2SnO_3
&c.		&c.

Valency of the Halogens.

These double fluorides, however, are not more numerous or prominent than are the double chlorides, bromides, and iodides, and it seems not unreasonable to explain their existence by a similar hypothesis. The atom $F\,(=19)$ may be occasionally trivalent. If so, the constitution of fluoride of potassium and hydrogen may be represented as

$$KF = FH,$$

and that of the other fluorides, single and double, in a similar manner.

Valency of the Halogens.—An atom of a halogen never replaces, in a direct manner, more than one atom of hydrogen. In the haloid salts of these elements, 35·5 parts of chlorine and equivalent quantities of bromine, iodine, and fluorine are almost always combined with the metallic representative of one part by weight of hydrogen. The halogens are therefore generally univalent. Nevertheless, many compounds are known, the existence of which can scarcely be accounted for except upon the hypothesis of their occasional trivalent function.

Thus, in addition to the normal iodides and iodo-substitution compounds,

$$H'I, \quad K'I, \quad Hg''I_2, \quad C_2H_3I'O_2,$$

iodine forms the following compounds:

$$I'''Cl_3, \quad I'''(C_2H_3O_2)_3, \quad KAgI'''_2,$$

which may be written as follows:

$$\underset{Cl}{\overset{Cl}{I{-}Cl}} \quad \text{or} \quad \underset{Cl}{\overset{Cl}{I{-}Cl}} \quad I{-}\!\!\begin{array}{l}O-C_2H_3O\\O-C_2H_3O\\O-C_2H_3O\end{array} \quad K-I=I-Ag$$

In their oxygenated compounds, chlorine, bromine, and iodine present also a very marked resemblance to nitrogen, which is most usually (perhaps always) a triad. Thus we have hypochlorous and hyponitrous acids, both extremely unstable bodies, known chiefly in the form of their salts; chlorous and nitrous acids, also very unstable; chloric and nitric acids, both liquid, easily decomposable, highly corrosive bodies, the salts of which are all soluble in water. The nitrogen analogue of perchloric acid is at present unknown. The formulæ of these corresponding pairs of compounds are as follows:

$$HClO \qquad HClO_2 \qquad HClO_3 \qquad HClO_4$$
$$HNO \qquad HNO_2 \qquad HNO_3$$

A further correspondence is observable in their oxides:

$$Cl_2O \qquad Cl_2O_3\,(?) \qquad ClO_2 \qquad\qquad I_2O_5$$
$$N_2O \qquad N_2O_3\,(?) \qquad NO_2 \text{ and } N_2O_4 \qquad N_2O_5$$

Now, if we admit that nitrogen is trivalent in these compounds, the presumption that chlorine and its congeners are also trivalent is, at least, worthy of discussion. The following graphic formulæ express the constitution of chloric and nitric acids upon this hypothesis:—

$$H-O-Cl\!\!<\!\!\genfrac{}{}{0pt}{}{O}{O}\!\!| \qquad\qquad H-O-N\!\!<\!\!\genfrac{}{}{0pt}{}{O}{O}\!\!|$$

Chloric acid.   Nitric acid.

As contributing evidence in support of the same view, we might also point to the large number of permanent and definite double compounds which the chlorides, bromides, and iodides are capable of forming. We have, for example, the well-known chloro-platinates of the alkali metals—

$$2NaCl.PtCl_4$$
$$2KCl.PtCl_4, \&c.,$$

besides the innumerable compounds of the alkali-metal chlorides with the chlorides of magnesium, iron, aluminium, mercury, copper, and other metals. Although a few of these compounds are decomposed by water, the majority of them are stable enough, and give no signs of alteration by such treatment. Their constitution is most reasonably represented upon the assumption of the trivalence of the halogen. The compound $KI.HgI_2$, for example, may be written in this manner:

$$K-I\underset{I}{\overset{I}{\diamondsuit}}Hg$$

Some chemists regard these compounds as formed by the union of entire molecules of the constituent salts, combined together by some kind of adhesion differing from ordinary chemical affinity. Thus the potassio-mercuric iodide represented above is considered to contain both iodide of potassium and iodide of mercury, $HgI_2 + KI$. This salt, however, is colourless, soluble in water, and totally unlike the scarlet insoluble mercuric iodide, the elements of which it contains, and, so far as physical characters are concerned, there is nothing to distinguish such cases from examples of ordinary atomic combination.

NON-METALS.—CLASS 2.

$$\left. \begin{array}{lll} \text{Oxygen} & . \ . & O = 16 \\ \text{Sulphur} & . \ . & S = 32 \\ \text{Selenion} & . \ . & Se = 79 \\ \textit{Tellurium} & . \ . & Te = 126 \end{array} \right\} \frac{S + Te}{2} = 79.$$

These elements are associated together principally by reason of the correspondence in composition and general properties among their compounds with hydrogen—

$$OH_2 \quad SH_2 \quad SeH_2 \quad TeH_2.$$

For although, taking any two adjacent terms of the series, many points of resemblance may be traced out, yet there is a very wide interval between the colourless gas oxygen and the silvery, metallic, crystalline solid tellurium.

Oxygen, standing apart from the rest in virtue of many extraordinary qualities, deserves to be noticed first. This element constitutes nearly one-half the total weight of the earth's crust. Indeed, it seems to be the preponderating element of the globe.

Leaving out of account the insignificant quantities of metals and metallic sulphides embedded in the earth, it may be said that the entire constituents of the earth's crust, including the water of the ocean, consist of compounds saturated with oxygen; and even supposing all organic bodies to be completely burnt up, the atmosphere would still contain a considerable quantity of unemployed oxygen.

Oxygen is entirely unmatched among the rest of the elements, both as regards the number and varied character of its compounds, and the important part which it plays in relation to combustion and life.

Water, hydrogen protoxide, also exhibits properties which are in every way remarkable as specially fitting it for the part it has to play in the economy of nature—the very type of a neutral body, yet capable, under special circumstances, of acting either as a feeble base or a feeble acid; an almost universal solvent of saline bodies; exhibiting the anomaly of attaining maximum density at a temperature (4°) above its solidifying point; standing alone amongst liquids by reason of its great capacity for heat; having a vapour-density little more than half that of its gaseous sulphur analogue (SH_2), and yet liquefying and solidifying readily.

The higher oxide of hydrogen, O_2H_2, with its bleaching and oxidising powers, resembles in no slight measure the element chlorine.

The name *oxygen*, 'acid producer,' from ὀξύς and γεννάω, was given by Lavoisier under the mistaken impression that this element contained a principle common to all acids. This we know was an error. Nevertheless, the name was not ill-chosen, and for the following reasons:—

1. The majority of known acids contain oxygen. The chief exceptions to this statement are as follows:—

(a) The so-called hydracids, HF, HCl, HBr, HI, HCN, and their compounds, such as fluoboric acid, $HF.BF_3$ or HBF_4, fluosilicic acid, $2HF.SiF_4$ or H_2SiF_6, chlorauric and chloroplatinic acids, $HCl.AuCl_3$ or $HAuCl_4$, $2HCl.PtCl_4$ or H_2PtCl_6, &c.

(b) Thio-acids, which may be regarded as oxygen acids, a part or the whole of the oxygen of which is replaced by sulphur, e.g.—

Thiosulphuric acid . . . H_2SO_3S
Thiocarbonic acid . . . H_2CS_3
Thiacetic acid . . . HC_2H_3OS

2. In proportion as the quantity of oxygen increases in a series of acids containing the same elements, so the acidity as well as the stability of the compound is almost always increased. A detailed examination of such examples as the following would, if space permitted, confirm this statement:

$HClO$	HNO_2	H_2SO_3	H_3PO_2
$HClO_2$	HNO_3	H_2SO_4	H_3PO_3
$HClO_3$			H_3PO_4
$HClO_4$			

Glycolic acid, $C_2H_4O_3$
Oxalic acid, $C_2H_2O_4$

3. The addition of oxygen to (neutral) aldehyds converts them into acids. Ex. gr.—

Acetic.
Aldehyd . . . . C_2H_4O
Acid . . . . $C_2H_4O_2$

Benzoic.
Aldehyd . . . . C_7H_6O
Acid . . . . $C_7H_6O_2$

4. In homologous series of acids, the lowest terms which are richest in oxygen show a far stronger development of

the acid character than the higher terms of the same series, which are comparatively poor in oxygen. Compare, for example—

Formic acid, CH_2O_2, containing 69·5 per cent. of oxygen,
Acetic acid, $C_2H_4O_2$,, 53·3 ,, ,,
with
Valeric acid, $C_5H_{10}O_2$,, 31·3 ,, ,,
and
Palmitic acid, $C_{16}H_{32}O_2$,, 12·4 ,, ,,

5. Many oxides of metals are basic, that is, saturate acids with production of water and salt, but a few examples will be sufficient to show that it is only those which contain a comparatively small quantity of oxygen that possess this power. As the quantity of oxygen increases in a given series, the basic character gradually disappears, and gives place to a more or less decided acid-forming tendency.

EXAMPLES OF METALLIC OXIDES.

Basic	Hg_2O HgO	K_2O	BaO	PbO	MnO Mn_2O_3 (feebly)	CrO Cr_2O_3 (feebly)
Intermediate		K_2O_2 K_2O_4	BaO_2	PbO_2	MnO_2	
Anhydric Acids					MnO_3 Mn_2O_7	CrO_3

Allotropic oxygen or *ozone* is a body which has attracted considerable attention. The experiments of Brodie finally decided the question of its constitution in favour of the hypothesis long ago put forward by Odling. Ozone is now proved to be allotropic oxygen, free from hydrogen, and to have the formula O_3. It may thus be regarded as formed on the type of hydric peroxide, with which body it agrees in many of its reactions.

Sulphur, Selenion.—These two elements resemble each other closely. Sulphur is a yellow, selenion a red solid, exhibiting several modifications, some of which are crystalline and soluble in carbon disulphide, others amorphous and insoluble.

Principal Allotropes of Sulphur.

	Rel. Dens.	In carbon disulphide.
1. Octahedral (native)	2·07	soluble.
2. Prismatic (monoclinic)	1·98	transformed into 1.
3. Plastic	1·95	insoluble.
4. Amorphous	1·95	insoluble.

(Precipitated from sulphur chloride or from thiosulphates.)

Principal Allotropes of Selenion.

1. Monoclinic (native)	4·4	soluble.
2. Crystalline (form?)	4·8	insoluble.
3. Vitreous	4·3	insoluble.
4. Amorphous	?	insoluble.

(Precipitated from selenious acid.)

The following is a comparison of the most important compounds of sulphur and selenion :—

H_2S	SO_2	SO_3	H_2SO_3	H_2SO_4
Gas.	Gas.	Solid.	Crystallisable at low temperatures.	Oily liquid.
H_2Se	SeO_2	SeO_3	H_2SeO_3	H_2SeO_4
Gas.	Solid.	Unknown.	Crystalline solid.	Oily liquid.

The selenates are isomorphous with the sulphates.

The relations of sulphur to oxygen are shown in the following synopsis of some of their compounds :—

SULPHUR.	OXYGEN.
SH_2 gas, feebly acid S_2H_2 oily liquid CS red solid? CS_2 volatile liquid COS gas	OH_2 O_2H_2 oily liquid CO colourless gas CO_2 gas
ACIDS.	
$C_2H_3O.SH$ thiacetic. B.P. 93° $SO_2.OH.SH$ thiosulphuric	$C_2H_3O.OH$ acetic. B.P. 120° $SO_2.OH.OH$ sulphuric
BASES.	
SKH SK_2 SCa $CS(NH_2)_2$ thiocarbamide	OKH OK_2 OCa $CO(NH_2)_2$ carbamide or urea
SALTS.	
M'_2CS_3 thiocarbonate M'_3AsS_4 thioarsenate $M'CNS$ thiocyanate	M'_2CO_3 carbonate M'_3AsO_4 arsenate $M'CNO$ cyanate
ALCOHOLS.	
$S\{{C_2H_5 \atop H}$ mercaptan. B.P. 36°	$O\{{C_2H_5 \atop H}$ B.P. 78°·4
ETHERS.	
$S(C_2H_5)_2$ B.P. 72°	$O(C_2H_5)_2$ B.P. 35°·6

It is necessary, however, to add that many oxygen compounds are known for which at present there are no corresponding terms in the sulphur series—the oxides of nitrogen, for example. On the other hand, sulphur has a power of accumulating in a manner which is not exhibited, at least to the same extent, by oxygen. Consequently, many polysulphides exist for which there are no corresponding oxides. Thus we have K_2S_5, CaS_5, FeS_2, $(C_2H_5)_2S_2$, and $(C_2H_5)_2S_3$.

Valency of Oxygen and Sulphur.

Oxygen is generally diad or bivalent, and in gaseous compounds, with the exception of carbonic oxide, invariably so. But that it is capable of acting the part of a tetrad there can be no doubt, as in the oxides Cu_4O and Ag_4O. Carbonic oxide is generally represented as containing unsaturated carbon, $C = O$; but admitting the essentially tetrad character of the oxygen, we should represent it as constituted thus, $C \equiv O$. Again, water of crystallisation is almost universally considered by chemists to exist as water in the salts in which it occurs, the molecules of the salt and the molecules of the water retaining their individuality, being united only by some kind of adhesion. Such water is very easily detached, and constitutes no essential part of the chemically reacting unit of the compound in all the most characteristic of its transformations. Nevertheless, the idea that this water is held to the salt by the two extra units of combining power of the oxygen is quite worthy of consideration. One example of the application of this hypothesis will suffice. Zinc sulphate contains $ZnSO_4.7OH_2$, or $ZnSO_4OH_2.6OH_2$. This may be expressed in the following manner :—

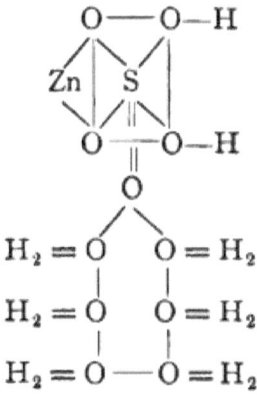

Now it is well known that six out of the seven molecules of water present in crystals of zinc sulphate are expelled by heat much more readily than the last molecule, which can be detached only with great difficulty. This fact is recorded in the second of the two formulæ given above, and it also comes out in the graphic formula in which two atoms of hydrogen are represented as occupying a position in the molecule corresponding with that of the zinc.

With regard to sulphur, the question is somewhat different. This element forms three well-marked classes of compounds. In the first of these it is bivalent, and may be regarded as the strict representative of oxygen. Examples of this type of compound have been already given, SH_2, S_2H_2, CS_2, COS.

In the second class the sulphur alone is quadrivalent: *ex. gr.* in—

$$\overset{iv}{S}\begin{cases}Cl\\Cl\\Cl\\Cl\end{cases} \qquad S\begin{cases}Cl\\Cl\\O''\end{cases} \text{ and } \quad S\begin{cases}O''\\O''\end{cases}$$

Sulphur tetrachloride. Thionyl chloride. Sulphur dioxide.

In the third class of sulphur compounds, represented by sulphuric anhydride, SO_3, the central atom seems to be sexivalent sulphur: $\overset{O}{\underset{O=S=O}{\|}}$. It is true that the constitution of this and all the rest of the sulphur compounds might be represented on the hypothesis that sulphur is diad,

$$\begin{matrix} O-S-O \\ \searrow\; \swarrow \\ O \end{matrix}$$

, but the analogy of sulphur with chromium, and especially the isomorphism of the sulphates and chromates, seem to establish the title of sulphur to be regarded as a hexad.

Sulphur Compounds. 235

The several oxides, hydrates, and oxychlorides are therefore represented in the following manner :—

$$\overset{vi}{S}\begin{cases} O'' \\ O'' \\ O'' \end{cases} \qquad \overset{vi}{S}\begin{cases} O'' \\ O'' \\ Cl \\ Cl \end{cases} \qquad \overset{vi}{S}\begin{cases} O'' \\ O'' \\ (OH)' \\ Cl \end{cases} \qquad \overset{vi}{S}\begin{cases} O'' \\ O'' \\ (OH)' \\ (OH)' \end{cases}$$

Sulphur trioxide. Sulphuryl chloride. Sulphuryl chlorhydrate. Sulphuryl hydrate, or common sulphuric acid.

$$\overset{vi}{S}\begin{cases} O'' \\ (OH)' \\ (OH)' \\ (OH)' \\ (OH)' \end{cases} \qquad \overset{vi}{S}\begin{cases} O \\ O \\ O \end{cases}\!\!>\!Fe \\ \begin{matrix} (OH)' \\ (OH)' \end{matrix}$$

Glacial or tetrahydric sulphuric acid. Ferrous sulphate (dried at 100°).

$$\overset{vi}{S}\begin{cases} OH \\ OH \\ OH \\ OH \\ OH \\ OH \end{cases} \qquad S\begin{cases} \overset{O}{\underset{O}{}}\!\!>\!Hg \\ \overset{O}{\underset{O}{}}\!\!>\!Hg \\ \overset{O}{\underset{O}{}}\!\!>\!Hg \end{cases}$$

Orthosulphuric acid (unknown). Trimercuric orthosulphate, Turpeth mineral.

$$\begin{cases} SO_2.OH \\ O \\ SO_2.OH \end{cases} \qquad \begin{cases} SO_2.Cl \\ O \\ SO_2.Cl \end{cases}$$

Pyrosulphuric or Nordhausen sulphuric acid. Pyrosulphuryl chloride.

NON-METALS.—CLASS 3 : BORON.

A triad element, but exhibiting marked analogies with tetrad silicon. The following are their chief points of agreement :—

Boron.	Silicon.
Element, (a) amorphous brown powder or (b) quadratic octahedrons	Element, (a) amorphous brown powder or (b) hexagonal plates
Density 2·68	Density about 2·5
BH_3 inflammable gas	SiH_4 spontaneously inflammable gas
$B(CH_3)_3$, $B(C_2H_5)_3$, &c.	$Si(CH_3)_4$, $Si(C_2H_5)_4$, &c.
$BiCl_3$. B.P. 17°	$SiCl_4$. B.P. 59°
BF_3 gas	SiF_4 gas
$BF_3.HF$ or BHF_4 known only in solution: salts crystallisable	$SiF_4.2HF$ or SiH_2F_6 scarcely known except in solution: salts tolerably stable
H_3BO_3 crystallisable	H_4SiO_4 exists only in solution
HBO_2	H_2SiO_3
B_2O_3 (fused, Dens. 1·83)	SiO_2 (fused, Dens. 2·2)
M'_3BO_3 orthoborates	M'_4SiO_4 orthosilicates
$M'BO_2$ metaborates	M'_2SiO_3 metasilicates
Many anhydroborates	Many anhydrosilicates
$M'BO_2 + xB_2O_3$	$M'_2SiO_3 + xSiO_2$
Ex. fused borax, $2NaBO_2 + B_2O_3$	Ex. orthoclase felspar $\left. \begin{array}{l} K_2SiO_4 \\ Al_2(SiO_3)_3 \end{array} \right\} + 2SiO_2$

Boron is one of the few elements which combine in a direct manner with nitrogen. Boron nitride has the formula BN, or perhaps B_3N_3.

NON-METALS.—CLASS 4.

Carbon, C = 12.
Silicon, Si = 28.

Allotropes of Carbon.

a. Diamond, octahedral, Dens. 3·3 to 3·5.
b. Graphite, hexagonal, Dens. about 2·2.
c. Charcoal, amorphous.

Allotropes of Silicon.

a. Adamantine (?).
b. Graphitic, hexagonal, Dens. about 2·5.
c. Amorphous.

Carbon is remarkable as being the essential element of organic nature, and silicon as being one of the most abundant constituents of the earth's solid crust. Regarded as chemical elements, they present problems of special interest in the multitudinous array of compounds which they have furnished to the chemist. The carbon compounds will be specially considered hereafter; it therefore only remains to point out in this place a few of the remarkable coincidences of composition so frequently observed between the compounds of these two elements. The number of bodies in which silicon replaces one or more atoms of the carbon in some already familiar compound is rapidly increasing, and still further developments in this direction may be looked for.

In comparing carbon and silicon compounds together, it cannot be said that the similarity so often noticed in their constitution extends to their properties. In most cases corresponding carbon and silicon compounds are very different bodies. A few examples only can be referred to:

Carbon Compounds.	Silicon Compounds.
CO, colourless inflammable gas.	SiO, unknown.
CO_2, colourless gas, soluble to some extent in water.	SiO_2, crystalline solid, fusible only at a very high temperature, totally insoluble in water. Hydrates slightly soluble.
CH_4, permanent gas.	SiH_4, spontaneously inflammable gas.
$CHCl_3$, heavy liquid, unacted upon by water.	$SiHCl_3$, liquid, decomposed by water.

Orthosilicic and orthocarbonic acids, H_4SiO_4 and H_4CO_4, are unknown, and but very few of their compounds have been studied. Of orthosilicates we have olivine, Mg''_2SiO_4, ferrous orthosilicate, Fe''_2SiO_4, the ethyl salt, $(C_2H_5)_4SiO_4$, and a few others. Orthocarbonates are represented by such compounds as the ammonium-hydrogen orthocarbonate $(NH_4)_2H_2CO_4$, and the ether $(C_2H_5)_4CO_4$. Ordinary carbonates are derived from metacarbonic acid, H_2CO_3, a body which has not been isolated, though the corresponding silicic acid has been obtained as a solid glassy mass. All carbonates are decomposed on the addition of even the weakest acids, whereas the majority of the silicates are attacked but very slowly by acids. Nevertheless silicates containing alkaline or earthy bases are slowly decomposed even by carbonic acid, which extracts the alkali, leaving hydrated silica, alumina and ferric oxide, &c.

Valency of Carbon and Silicon.

Silicon is always tetrad. Carbon is tetrad in all its known compounds, except possibly carbonic oxide, CO, and the isocyanides. The unsaturated character of carbonic oxide is shown by the readiness with which it enters into combination with chlorine, with oxygen, with potassium, with caustic potash, and with cuprous or platinous chloride.

$$CO + Cl_2 = COCl_2, \text{ carbonyl chloride}$$
$$CO + O = CO_2, \text{ carbon dioxide}$$
$$nCO + K_n = K_nC_nO_n$$
$$CO + KHO = KCHO_2, \text{ formate, &c.}$$

From the characteristic reactions of the isocyanides (see Isomerism, Chap. XXI., p. 197) it appears that in these bodies an atom of carbon is linked to the hydrogen, or other positive radicle present, by the atom of nitrogen. Now, according as we regard the nitrogen in these compounds as discharging a triad or pentad function, we must attribute to the carbon a replacing value equal to two or four atoms of hydrogen. Either of these two formulæ, then, may be adopted :—

$$\overset{\prime\prime}{C} = \overset{\prime\prime\prime}{N} - \overset{\prime}{R} \quad \text{or} \quad \overset{\text{iv}}{C} \equiv \overset{\text{v}}{N} - \overset{\prime}{R}.$$

The one may be regarded as a derivative of ammonia, the other of marsh-gas.

In favour of the former view we may adduce the combination of such a cyanide as prussic acid with hydrochloric,

Nitrogen and Phosphorus.

hydrobromic, and hydriodic acids, forming crystalline compounds analogous to sal-ammoniac, and with hydrogen forming methylamine ; also the combination of the isocyanides generally with oxygen to form isocyanates, and with sulphur to form isothiocyanates. These combinations may be represented thus :—

$$N\begin{cases}H\\C''\end{cases} \quad N\begin{cases}H\\C''\\HCl\end{cases} \quad N\begin{cases}H\\C=O\end{cases} \quad N\begin{cases}H\\C=S\end{cases} \quad N\begin{cases}H_2\\CH_3\end{cases}$$

Hydrocyanic acid or Carbylamine.   Carbylamine hydrochloride.   Cyanic acid.   Thiocyanic acid.   Methylamine.

If the second hypothesis is adopted, we must imagine that the union of the carbon to the nitrogen is partly unlocked before these compounds can be found. Written on the marsh-gas type the formulæ then become—

$$C^{iv}(NH)^{iv} \quad C\begin{cases}HCl\\(NH)''\end{cases} \quad C\begin{cases}O''\\(NH)''\end{cases} \quad C\begin{cases}S''\\(NH)''\end{cases} \quad C\begin{cases}H_3\\(NH_2)'\end{cases}$$

NON-METALS.—CLASS 5 : NITROGEN AND PHOSPHORUS.

$$\left.\begin{array}{ll}\text{Nitrogen} & N=14\\ \text{Phosphorus} & P=31\\ \text{Vanadium} & V=51\end{array}\right\} \frac{N+V}{2} = 32\cdot5 \quad \left.\begin{array}{ll}\text{Phosphorus} & P=31\\ \text{Arsenic} & As=75\\ \text{Antimony} & Sb=120\end{array}\right\} \frac{P+Sb}{2} = 75\cdot5$$

$$\left.\begin{array}{ll}\text{Phosphorus} & P=31\\ \text{Vanadium} & V=51\\ \text{Arsenic} & As=75\end{array}\right\} \frac{P+As}{2} = 53 \quad \left.\begin{array}{ll}\text{Phosphorus} & P=31\\ \text{Antimony} & Sb=120\\ \text{Bismuth} & Bi=208\end{array}\right\} \frac{P+Bi}{2} = 119\cdot5$$

Diagram of Numerical Relations typical of Chemical Relations.

$$\begin{array}{c}N14\\|\quad\diagup\; V51-As75\\P31\!-\!\!-\!As75-Sb120\\|\quad\diagdown\\V51\quad Sb120-Bi208\end{array}$$

Nitrogen is a colourless gas, very slightly soluble in water, liquefiable with great difficulty, and not known in any allotropic form.

Phosphorus, on the contrary, is, in its ordinary condition, a crystalline solid, though capable of passing into several allotropic modifications. These have already been referred to (Chap. XXI., p. 199).

The relative density of nitrogen agrees with the molecular formula N_2, whilst the vapour density of phosphorus corresponds with P_4. This peculiarity of phosphorus is shared by the semi-metal arsenic.

Both nitrogen and phosphorus are intimately concerned in the processes of animal life, phosphorus being especially abundant in the form of phosphates in bone, and in nervous tissue.

Nitrogen is connected with phosphorus mainly by reason of the resemblance between ammonia, NH_3, and phosphine, PH_3, and the bases formed from them by substitution of hydrocarbon radicles for the hydrogen. The element itself, as a permanent incombustible gas, is wholly unlike the solid inflammable phosphorus, whilst its oxides and acids, in their volatility and easy decomposability, form a strong contrast to the corresponding compounds of phosphorus. The accident of agreement in their crude or synoptic formulæ by no means implies identity of constitution, and, for the reasons just alluded to, it seems not improbable that the atomic structure of the nitrogen oxides and acids is, at least in many cases, different from that of the phosphorus compounds.

The following are almost the only known examples of concordance in the crude formulæ of the oxygen compounds :—

Nitrogen trioxide.	Phosphorus trioxide.
N_2O_3	P_2O_3
Tetroxide.	Tetroxide.
N_2O_4	P_2O_4
Pentoxide.	Pentoxide.
N_2O_5	P_2O_5
Nitric acid.	Metaphosphoric acid.
HNO_3	HPO_3

The compounds N_2O, NO, HNO_2, find no parallels in the phosphorus series, whilst H_3PO_4, H_3PO_3, H_3PO_2, $H_4P_2O_7$, the salts of which, at least, are highly stable and definite, are altogether unrepresented among the nitrogen compounds. Again, whilst phosphorus yields two chlorides, PCl_3 and PCl_5, and an oxychloride, $POCl_3$, nitrogen forms no chloride,[1] but an oxychloride, $NOCl$, a yellow gas. It has been the very general custom to represent all the compounds of nitrogen by formulæ framed in the same manner as those of the phosphoric compounds, without regard to their dissimilar properties. But by assuming that nitrogen is trivalent in these bodies we render some account of these discrepancies. On this view the formulæ given above for nitric acid and nitric anhydride may be expanded in the following manner :[2]—

Nitric pentoxide.

$$\begin{matrix} O \\ | \\ O \end{matrix} \!\!\!> N\!-\!O\!-\!N <\!\!\!\begin{matrix} O \\ | \\ O \end{matrix}$$

Phosphoric pentoxide.

$$O\!=\!P\!-\!O\!-\!P\!=\!O$$
$$\quad\;\|\qquad\quad\;\|$$
$$\quad\;O\qquad\;\;O$$

Nitric acid.

$$\begin{matrix} O \\ | \\ O \end{matrix}\!\!\!> N\!-\!O\!-\!H$$

Metaphosphoric acid.

$$O\!=\!P\!-\!O\!-\!H$$
$$\quad\;\|$$
$$\quad\;O$$

In the ammonium compounds, on the other hand (p. 308), nitrogen is probably pentad—chloride of ammonium, for instance, being N^vH_4Cl.

Phosphorus.—The two chlorides PCl_3 and PCl_5 may be taken as the types of the two chief classes of compounds which phosphorus is capable of forming.

[1] The so-called chloride of nitrogen, obtained by the action of chlorine upon sal-ammoniac, probably contains hydrogen. Its formula is supposed to be $NHCl_2$, but in consequence of its dangerously explosive properties it has been only imperfectly examined.

[2] The formulæ might also be written $O=N-O-O-O-N=O$, and $O=N-O-O-H$, though for several reasons the form given in the text is preferable.

The trichloride PCl_3 is a colourless volatile liquid, which combines in a direct manner with chlorine, with oxygen, or with sulphur, yielding the pentachloride, PCl_5, the oxychloride, PCl_3O, and the sulphochloride, PCl_3S, in which compounds the phosphorus appears to be saturated.

The best known iodide is anomalous. It has the formula P_2I_4, and corresponds with a liquid hydrogen compound P_2H_4, to which the spontaneous inflammability of phosphoretted hydrogen is stated to be due.

Phosphine, or phosphoretted hydrogen, PH_3, is a highly inflammable gas, which so far imitates ammonia[1] that it combines with hydriodic acid to form a salt, phosphonium iodide, PH_4I, crystallisable in cubes.

The compounds in which phosphorus plays the part of a pentad are more numerous. Besides the pentachloride PCl_5, and pentoxide P_2O_5, and intermediate PCl_3O, we have the following series of oxidised bodies which may be considered as derived from phosphine, either by addition of oxygen or by substitution of hydroxyl for its hydrogen:—

Phosphine.

H_3P or

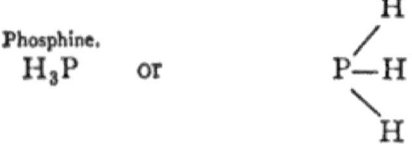

Phosphine oxide.

H_3PO, unknown, but represented by—

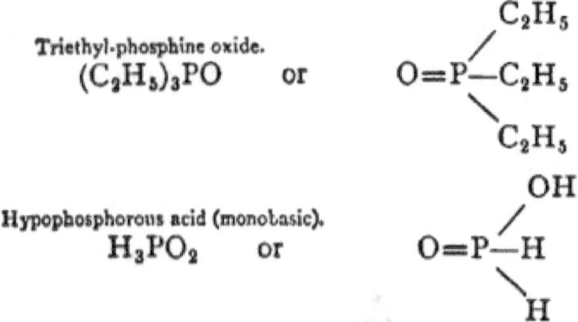

Triethyl-phosphine oxide.

$(C_2H_5)_3PO$ or

Hypophosphorous acid (monobasic).

H_3PO_2 or

---

[1] See also Amines, p. 310.

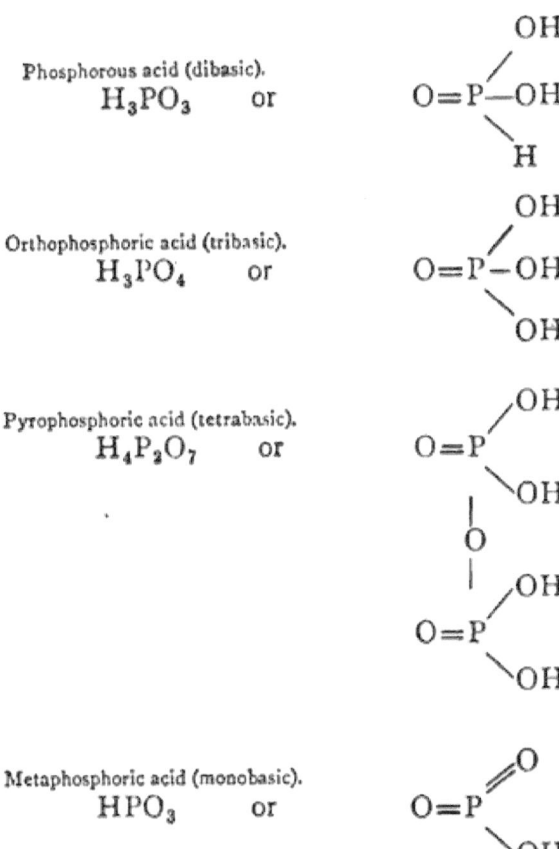

With the exception of the first two compounds in the list, phosphine and its oxide, these bodies are all powerful acids, syrupy, glassy, or crystalline, according to circumstances. They cannot be made to evolve oxygen at any temperature as nitrous and nitric acids do. Under the influence of heat, hypophosphorous and phosphorous acids emit phosphine, leaving a residue of phosphoric acid—

$$2H_3PO_2 = H_3PO_4 + H_3P$$
$$\text{and } 4H_3PO_3 = 3H_3PO_4 + H_3P;$$

whilst orthophosphoric acid by the same treatment loses

water, though the dehydration can by no means be carried beyond the production of metaphosphoric acid:

$$2H_3PO_4 - H_2O = H_4P_2O_7$$
$$\text{and } H_3PO_4 - H_2O = HPO_3$$
$$\text{or } H_4P_2O_7 - H_2O = 2HPO_3$$

By boiling either of these products with water the ortho-acid is re-generated.

Metaphosphoric acid is convertible, by the action of heat, into several polymeric acids, $H_n P_n O_{3n}$.

Phosphorus is closely connected with the semi-metals vanadium and arsenic, by reason of the very general isomorphism of the phosphates with the arsenates and vanadates. This relationship will be adverted to in a subsequent chapter.

CHAPTER XXIV.

Division II.—Metalloids.

The name *metalloid* obviously belongs to bodies which have the form or appearance of a metal, or which in some way resemble a metal. Hence the impropriety of applying this term, as is not uncommonly done, to such elements as the halogens, to nitrogen, or phosphorus. The name is intended in this volume to designate a body which, although resembling a metal in most characteristics, yet lacks some one or more of the features which true metals generally present. It applies to a somewhat miscellaneous set of bodies. The question of the propriety of including hydrogen amongst them will be discussed presently, but, leaving this one element out of consideration, the test by which the sub-metallic character of most of them may be detected is the formation of more or less definite oxy-salts. Their oxides are for the most part well-marked anhydric acids, but each of them is capable of producing at least one oxide which possesses basic tendencies more or less pronounced. Still, putting hydrogen aside, the metalloids are comparatively imperfect conductors of heat and electricity, and generally brittle. Tin is, however, both malleable and moderately ductile.

METALLOIDS.—I. HYDROGEN.

Hydrogen is isolated from all other known bodies by reason of its extreme lightness, its relative density being only ·0693 when air is 1. Chemically it differs from all the non-metals in manifesting no tendency to combine with metals; only one such compound, cuprous hydride, Cu_2H_2, being known.[1] On the other hand, hydrogen is, of all the elements, most ready to lend itself to that peculiar state of mechanical combination with metals known as 'occlusion' (Chap. VI., p. 44). Palladium, for example, absorbs hydrogen largely, and the charged metal is indistinguishable in appearance from pure palladium, though its conducting power is slightly diminished. This body was regarded by Graham, who discovered it, as the realisation of the long-cherished idea of the essentially metallic character of hydrogen. But its title to recognition as a metal or metalloid is based more securely upon the resemblance which acids or hydrogen salts bear to metallic salts. In spite of a considerable number of gaseous and liquid exceptions, the majority of known acids are crystallisable bodies, often combining with water of crystallisation, and utterly indistinguishable to all appearance from ordinary saline compounds. Among common acids, for example, the following are crystallisable:—

Sulphuric acid . . H_2SO_4 (below $-35°$).
Glacial sulphuric acid . H_4SO_5 (below $9°$).
Phosphoric acid . . H_3PO_4.

[1] Definite compounds of hydrogen with palladium, sodium, and potassium, having the formulæ Pd_2H, Na_2H, and K_2H, are said to be produced by heating these metals in hydrogen gas. It is, however, not yet certain that these are true chemical compounds.

Boracic acid . . . H_3BO_3.
Oxalic acid . . . $H_2C_2O_4 + 2H_2O$.
Formic acid . . . $HCHO_2$ (below 1°).
Acetic acid . . . $HC_2H_3O_2$ (below 17°).
Tartaric acid . . $H_2C_4H_4O_6$.
Citric acid . . . $H_3C_6H_5O_7 + H_2O$.
Gallic acid . . . $HC_7H_5O_5 + H_2O$.

But the resemblance between hydrogen salts and metallic salts is not limited to external form. They agree in chemical reactions. When an acid is acted upon by a metal, and hydrogen is expelled, the metathesis is entirely comparable with the exchange of a more positive for a less positive metal of the ordinary recognised class. For example, zinc decomposes hydrogen sulphate, just as it decomposes copper sulphate; in each case the less positive metal is displaced, and a new salt formed.

$$\begin{cases} H_2SO_4 + Zn = H_2 + ZnSO_4. \\ CuSO_4 + Zn = Cu + ZnSO_4. \end{cases}$$

Again, water—that is, hydrogen oxide or hydroxide, H_2O or $H(HO)$—not unfrequently imitates metallic oxides or hydroxides in their chemical reactions. In order to produce exactly the same effect in any given case, it is only necessary to use the water in larger quantity or at a higher temperature.

The following are examples of this mode of action:—

$MgCl_2 + HgO = MgO + HgCl_2$.
$MgCl_2 + H_2O = MgO + 2HCl$.
$\begin{cases} Fe_2Cl_6 + 6NaHO = Fe_2(HO)_6 + 6NaCl. \\ Fe_2Cl_6 + 6HHO = Fe_2(HO)_6 + 6HCl. \end{cases}$

$\begin{cases} (C_3H_5,'''\overline{St}_3)^1 + 3NaHO = C_3H_5(HO)_3 + 3Na\overline{St}. \\ \quad\text{Stearine.} \qquad\qquad\qquad\qquad \text{Glycerine.} \qquad \text{Sodium stearate.} \\ (C_3H_5)'''\overline{St}_3 + 3HHO = C_3H_5(HO)_3 + 3H\overline{St}. \\ \quad\text{Stearine.} \qquad\qquad\qquad\qquad \text{Glycerine.} \quad \text{Hydrogen stearate, or stearic acid.} \end{cases}$

[1] $\overline{St} = C_{18}H_{35}O_2$.

Water and acids also undergo electrolysis in the same manner as metallic salts ; *e.g.*

$$2HHO = H_2 + (HO)_2\,^1$$
$$2HCl = H_2 + Cl_2$$
$$CuCl_2 = Cu + Cl_2$$

Lastly, it might be pointed out that the sour taste of hydrogen salts and their reddening effect upon vegetable blues are no more to be regarded as indications of peculiar constitution than is the alkaline reaction of many normal metallic salts or the neutrality of others. In such a series as the following we see a gradual transition from strongly alkaline to strongly acid properties :

Na_3PO_4	Na_2HPO_4	NaH_2PO_4	H_3PO_4
Very alkaline.	Alkaline.	Acid.	Very acid.

Enough has been said to indicate to the student that, despite many peculiarities which are in all probability connected with its remarkable physical characters, hydrogen displays, from a chemical point of view, features bearing an unmistakable metallic impress.

METALLOIDS.—2. TELLURIUM.

Tellurium in many respects closely imitates sulphur and selenion. Thus it forms a gaseous telluretted hydrogen, H_2Te, the representative of hydrogen sulphide ; also oxides TeO_2 and TeO_3, and acids H_2TeO_3 and H_2TeO_4, corresponding respectively with sulphur dioxide and trioxide, and with sulphurous and sulphuric acids. But tellurium is decidedly metallic in appearance, is thrown down from solution in the form of sulphide by sulphuretted hydrogen,

---

[1] It is a question whether *pure* water is an electrolyte at all, but supposing it to be capable of electrolysis this is probably the first phase in the decomposition, the oxygen which is evolved being due to a secondary reaction, thus :

$$2HHO = H_2 + (HO)_2 = H_2 + H_2O + O.$$

and its lower oxide saturates acids forming salts, of which the sulphate, $Te(SO_4)_2$, and nitrate, $Te(NO_3)_4$, are sufficient examples.

METALLOIDS.—3. GERMANIUM, TIN, TITANIUM, ZIRCONIUM.

Save for the distinctly metallic character of elemental tin and germanium, and production by tin, germanium, and titanium of an inferior diadic class of derivatives, these elements bear a strong family resemblance to silicon. That the metallic function is more or less distinctly developed in them, whilst it is altogether imperceptible in silicon, is also indicated by the existence of various oxy-salts, sulphates, phosphates, &c., of tin, titanium, and zirconium, and probably of germanium, similar compounds of silicon being entirely unknown.

The most characteristic compounds of these elements are the chlorides, oxides, and acids in which they play the part of tetrads. The following are some examples compared with corresponding silicon compounds:—

Silicon $Si = 28$	Germanium $Ge = 72$	Tin $Sn = 118$	Titanium $Ti = 48$	Zirconium $Zr = 89\cdot6$
$SiCl_4$ liquid, B.P. $50°$	$GeCl_4$ liquid, B.P. $87°$	$SnCl_4$ liquid, B.P. $120°$	$TiCl_4$ liquid, B.P. $135°$	$ZrCl_4$ volatile solid
SiF_4 liquefiable gas	GeF_4 solid?	SnF_4 fuming liquid?	TiF_4 fuming liquid	ZrF_4 solid
M'_2SiF_6 Fluosilicates	M'_2GeF_6 Fluo- germanates	M'_2SnF_6 Fluostannates	M'_2TiF_6 Fluotitanates	M'_2ZrF_6 Fluozirconates

Corresponding salts isomorphous.

SiO_2	GeO_2	SnO_2	TiO_2	ZrO_2
		in tinstone.	Isomorphous in rutile.	artificial.

Valency of Tin and Allied Elements.

As already mentioned, tin forms, in addition to the well-defined tetrad compounds, some of which are represented in the foregoing table, an oxide, Sn_nO_n, a chloride, Sn_nCl_{2n}, a sulphide, Sn_nS_n, and some other compounds in which it seems to be diad. In the absence of direct evidence in favour of the one view or the other, it is customary to represent the oxide as SnO, and the chloride as $SnCl_2$. In favour of the formula SnO we have its analogy to that of carbonic oxide, CO, though there is absolutely no resemblance in the properties of the two bodies to support such a view.

For stannous chloride, representing it as a dichloride, we have no example among the compounds of carbon and silicon. On the whole it seems, therefore, probable that the tin retains its quadrivalence in both the oxide and the chloride, and their formulæ in accordance with this view would be written thus :—

$$Sn_2Cl_4 \text{ or } Cl_2 = Sn = Sn = Cl_2$$
$$Sn_2O_2 \text{ or } O = Sn = Sn = O.$$

The element germanium was discovered in 1886 in a rare mineral, argyrodite, $3Ag_2S.GeS_2$, found at Freiberg. It is brittle and volatile, and crystallises in octahedrons of density 5·47, and melting at about 900°. It forms two classes of compounds somewhat similar to those of tin, including a yellow monosulphide, GeS, and a white disulphide, GeS_2.

It also forms a tetrethide, $Ge(C_2H_5)_4$, which boils at 160°, but no compound with hydrogen has yet been obtained.

METALLOIDS.—4. VANADIUM, ARSENIC, ANTIMONY, BISMUTH.

(For numerical relations of atomic weights see *Non-metals, Class* 5, p. 239.)

These elements are all quinquivalent in their most advanced states of combination, the tendency to assume this condition being the least prominent in the case of bismuth, which is of the whole class the most decidedly metallic.

Vanadium is remarkable for the extensive series of oxides, chlorides, and oxychlorides to which it gives rise, the different stages of oxidation of its dissolved compounds being characterised by a strange variety of colour, a yellow or red tint belonging to the higher, and a blue or green tint especially to the lower oxides, which thus resemble some of the compounds of chromium, molybdenum, and tungsten. Oxides of phosphorus, arsenic, antimony, and bismuth, except Bi_2O_5, are white.

The known oxides of vanadium have the formulæ V_2O, VO or V_2O_2, V_2O_3, VO_2 or V_2O_4, and V_2O_5, thus running parallel with the series of nitrogen oxides. There are three chlorides, VCl_2 or V_2Cl_4, VCl_3, and VCl_4. The last is a brown liquid boiling at 154°, and yielding a vapour of density 96·5, which is half the molecular weight represented by the formula VCl_4. Vanadium, therefore, does not maintain the usual persistence of odd or even valency throughout its combinations (see Chap. XVI.).

The vanadates, M'_3VO_4, are isomorphous with the phosphates, M'_3PO_4. Metavanadates and pyrovanadates also exist, corresponding with meta- and pyro-phosphates, besides a certain number of anhydro-meta-salts, $M''(VO_3)_2.V_2O_5$ and $M''(VO_3)_2.2V_2O_5$.

The analogies of arsenic, antimony, and bismuth will be most satisfactorily brought out by tabulating the formulæ of their leading compounds. For the sake of comparison the corresponding compounds of phosphorus are placed side by side with them.

The vapour densities of phosphorus and arsenic lead to the molecular formulæ P_4 and As_4 for those elements; but antimony and bismuth, though easily fusible metals, are not volatile enough to give vapour at any manageable temperature. Their molecular formulæ are therefore unknown.

Phosphorus	Arsenic	Antimony	Bismuth
Hydrides and Ethylides.			
PH_3	AsH_3	SbH_3	—
$P(C_2H_5)_3$	$As(C_2H_5)_3$	$Sb(C_2H_5)_3$	$Bi(C_2H_5)_3$
Chlorides.			
PCl_3	$AsCl_3$	$SbCl_3$	$BiCl_3$
PCl_5	—	$SbCl_5$	—
Oxides.			
P_2O_3	$(As_2O_3)_2$	$(Sb_2O_3)_2$	Bi_2O_3
P_2O_4		Sb_2O_4	Bi_2O_4
P_2O_5	As_2O_5	Sb_2O_5	Bi_2O_5
Acids.			
HPH_2O_2	—	—	—
—	$HAsO_2$?	$HSbO_2$	$HBiO_2$
H_2PHO_3	H_3AsO_3	—	
HPO_3	$HAsO_3$	$HSbO_3$	$HBiO_3$
H_3PO_4	H_3AsO_4	H_3SbO_4?	
$H_4P_2O_7$	$H_4As_2O_7$	$H_4Sb_2O_7$	
Sulphides.			
P_2S	—	—	—
—	As_2S_2	—	Bi_2S_2
P_2S_3	As_2S_3	Sb_2S_3	Bi_2S_3
P_2S_5	As_2S_5	Sb_2S_5	—

A few peculiarities exhibited by particular compounds occurring in this table demand notice.

1. The faculty of uniting with acids possessed in so remarkable a manner by ammonia is displayed by phosphoretted hydrogen towards hydriodic acid only, the corresponding compounds with hydrobromic and hydrochloric acids being produced only by compressing the mixed gases. Arsenic and antimony hydrides do not combine with acids.

2. The ethylated and corresponding methylated compounds are characterised not so much by a power of combining with acids as by a tendency to unite with O, S, Cl_2, I_2, &c., whereby the central element is brought to a state of saturation.

3. No pentachloride, pentabromide, oxychloride, or oxybromide of arsenic is known.

4. The vapour density of arsenious oxide (white arsenic) is 198. Hence its molecular weight is $198 \times 2 = 396$, and formula As_4O_6. It is usual to represent the oxides of phosphorus by the formulæ written in the table; but if the trioxide could be obtained in a pure state and its vapour density determined, it is by no means improbable that its molecular formula would also be doubled.

White arsenic occurs in commerce in the form of white masses with conchoidal fracture, often exhibiting a banded appearance, due to the presence of the isomeric vitreous and porcellanous varieties. The former is converted into the latter by pulverisation or by protracted boiling with water. Either variety is but slightly soluble in water, and the solution gives only a feebly acid reaction. Porcellanous arsenic probably owes its opacity to crystalline structure. By careful sublimation at regulated temperatures white arsenic may be obtained either in the form of transparent regular octahedra or in prismatic crystals of the trimetric system. It is, therefore, isodimorphous with antimonious oxide.

White arsenic is a feebly basic, as well as acid, anhydride. Its best known salt is the double tartrate, $K(AsO)C_4H_4O_6$, which is analogous to the potassio-tartrates of antimony, $K(SbO)C_4H_4O_6$ (tartar emetic), and boron, $K(BO)C_4H_4O_6$.

5. **Phosphoric oxide** is volatile and stable, but As_2O_5, Sb_2O_5, Bi_2O_5, are resolved by heat into oxygen and a lower oxide.

6. Arsenious acid appears to be normally tribasic, the yellow silver salt containing Ag_3AsO_3. It therefore differs from phosphorous acid, **which** is dibasic.

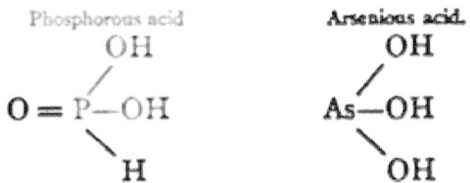

7. **The** orthophosphates and arsenates correspond in every respect with each other, and in many cases with the orthovanadates. In addition to the ordinary salts of the common phosphate type, $M'H_2PO_4$, M'_2HPO_4, and M'_3PO_4, there is a remarkable **series** of compounds occurring as minerals in combination with a chloride or fluoride.

Apatite $3[Ca_3(PO_4)_2] \cdot CaF_2$ or $Ca_5(PO_4)_3F$
      or $3[Ca_3(PO_4)_2] \cdot CaCl_2$ or $Ca_5(PO_4)_3Cl$
Pyromorphite $3[Pb_3(PO_4)_2] \cdot PbCl_2$ or $Pb_5(PO_4)_3Cl$
Mimetine $3[Pb_3(AsO_4)_2] \cdot PbCl_2$ or $Pb_5(AsO_4)_3Cl$
Vanadite $3[Pb_3(VO_4)_2] \cdot PbCl_2$ or $Pb_5(VO_4)_3Cl$.

Orthoantimonates and orthobismuthates are unknown, unless we consider the intermediate tetroxides in that light: antimonious orthoantimonate, Sb_2O_4 or $Sb'''(Sb^vO_4)$, and bismuthous orthobismuthate, Bi_2O_4 or $Bi'''(Bi^vO_4)$. Several orthothioantimonates are, however, known; the sodium salt, for example, $Na_3SbS_4 \cdot 9H_2O$, crystallising readily.

8. The sulphides of arsenic, antimony, and bismuth act as sulphur anhydrides, and in combination with the sulphides **of** silver, **lead, copper,** and other metals constitute several interesting minerals.

METALLOIDS.—5. NIOBIUM, TANTALUM.

These two rare metals may be regarded as about equally related on the one hand to tetrad tin and its allies, and on the other to hexad molybdenum and tungsten. With the latter elements, indeed, there is not only considerable analogy of properties, but a curious approximation of atomic weights.

$$Mo^{vi}\ 96 \qquad Nb^{v}\ 94$$
$$W^{vi}\ 184 \qquad Ta^{v}\ 182$$

Niobium and tantalum are pentad elements, and are characterised by the production of pentachlorides in the form of yellow, fusible, volatile bodies, the vapour densities of which accord with the formulæ $NbCl_5$ and $TaCl_5$ respectively.

The metals themselves are only known in the condition of black powders, which resist the attacks of most acids, except hydrofluoric acid, in which they are slowly soluble with evolution of hydrogen. Their fluorides unite with the fluorides of the alkali-metals, forming soluble and crystallisable double salts, $2KF,NbF_5$ and $2KF,TaF_5$. But niobium, at least, is also specially characterised by the production of oxyfluorides, which, in combination with the fluorides of the more positive metals, form salts which are isomorphous with the fluotitanates, fluostannates, fluotungstates, and fluozirconates. The following pairs of compounds, for example, crystallise in the same form:—

$$3KF \cdot HF \cdot NbOF_3$$
$$3KF \cdot HF \cdot SnF_4$$
$$2NH_4F \cdot NbOF_3$$
$$2NH_4F \cdot WO_2F_2$$
$$3NH_4F \cdot NbOF_3$$
$$3NH_4F \cdot ZrF_4$$
$$2NH_4F \cdot CuWO_2F_4$$
$$2NH_4F \cdot CuTiF_6$$

It thus appears that groups of elements of different valencies may occasionally replace one another isomorphously, provided the valency or atomicity of the various groups is the same.

The radicles which are thus at the same time chemically and isomorphously equivalent to one another in the foregoing compounds are as follows :—

$(TiF_2)''$, $(ZrF_2)''$, $(SnF_2)''$, $(NbOF)''$, and $(WO_2)''$.

Such relations are of precisely the same character as are those of the group $(NH_4)'$, ammonium, to the single metallic atom K', and precisely the same conclusions may be deduced from them.

METALLOIDS.—6.

Molybdenum	Mo =	95·7
Tungsten	W =	184
Uranium	U =	239.

These elements are heavily metallic bodies, tungsten and uranium being especially distinguished both by high atomic weight and specific gravity.

They appear to be hexads, though, of the three, tungsten alone produces a hexchloride, and that unstable. In the formation of trioxides and volatile dioxydichlorides (general formula $M^{vi}O_2Cl_2$) they resemble chromium, but they differ from that metal in furnishing neither sesquioxide nor monoxide analogous to Cr_2O_3 and CrO.

The infraction of the law of even numbers occurring in the case of the chlorides and bromides of tungsten has been already adverted to (Chap. XVI.). Molybdenum and uranium, like the rest, seem to form a pentachloride.

The trioxides, MoO_3, WO_3, and UO_3, are generally acid anhydrides. Some of the tungstates of the form M'_2WO_4 seem to be isomorphous with the corresponding chromates and sulphates, but the majority are extremely complex.

These trioxides also exhibit a feebly basic function, the oxy-salts of uranium being the most stable. The nitrate, for example, a yellow crystallisable salt, has the formula $(UO_2)''(NO_3)_2.6H_2O$, and the sulphate and oxalate are formed upon the same type.

Molybdenum and tungsten agree in forming trisulphides, MoS_3 and WS_3, which are unrepresented among the compounds of uranium and of chromium.

CHAPTER XXV.

Division III.—Metals.

METALS are all solids at the temperature of the air, with the sole exception of mercury, which is ordinarily liquid, but solidifies at $-40°$. The solid metals differ very much in their fusibility, the metals of the alkalies melting most readily (potassium at $62°\cdot5$), whilst platinum and its congeners require the highest temperature of the oxyhydrogen flame.

Metals, when in a compact state, combine great opacity with high reflective power, exhibiting the appearance which is sufficiently well known as metallic lustre—a property which, however, is exhibited by many other substances, such as iodine, graphite, galena, and many sulphides. Gold in very thin leaves transmits a greenish light.

That all the common metals are heavy is a fact familiar to every one; but an inspection of the table of densities given at the commencement of the section will lead to the conclusion that there are many which are much lighter than the familiar iron, copper, lead, and silver, whilst there are several which are even heavier than gold. As a curious fact, the heaviest solid known is a metal, platinum (rel. dens. $21\cdot5$), and the lightest known *solid* is also a metal, namely, lithium, whose relative density is only $\cdot578$.

But it is of much greater importance to observe that in many cases groups of metals which are associated together by reason of community of chemical properties, have also densities which nearly approximate to one another.

As examples of this may be cited the alkali metals, lithium, sodium, potassium, rubidium, cæsium ; the alkaline earth metals, magnesium, calcium, strontium, barium; the metals of the iron family, chromium, iron, manganese, nickel, cobalt, and copper ; silver, lead, and thallium ; and lastly the noble metals, gold, iridium, and platinum. Possibly other similar associations might be discovered, but for the fact that several metals are scarcely known in a compact form, and in other cases the numbers may be more or less incorrect in consequence of the presence of impurity in the specimens operated upon.

The following are the chief characteristics of metals as a class :—

1. Metals are the best conductors of heat. They differ widely among themselves in this respect, as the following rough comparison shows. Silver, as the best conductor, is placed at the head of the list. The numbers may be taken to represent the relative lengths of bars of equal diameter which, by applying a common source of heat to one extremity, would become equally heated in the same time.

Silver	1000
Copper	736
Gold	532
Iron	119
Lead	85
Platinum	84
Bismuth (an imperfect metal)	18

2. Metals are the best conductors of electricity. The metals given in the foregoing list stand in nearly the same order as regards electric conductivity.

3. Metals are almost always malleable, though in a few

S

cases—zinc, for example—this happens only at slightly elevated temperatures. Frozen mercury is malleable.

4. Many metals are ductile, but, since ductility is so largely dependent upon tenacity, it by no means follows that ductility should be manifested by a given metal in the same degree as malleability. Gold, for example, is by far the most malleable metal, though in point of ductility it is surpassed by platinum. Lead and tin, though easily rolled into sheet, can scarcely be obtained in the form of wire, owing to their very slight tenacity.

5. The oxides of metals are very generally basic (see p. 303.)

The basigenic character belongs exclusively to the metals and metalloids. But the oxygenic function is discharged almost equally well by particular elements in all three divisions.

The transition from metal to non-metal is, therefore, not accomplished by any sudden break, and the student must be prepared to encounter great, and perhaps insurmountable difficulties in any attempt to establish a line of demarcation between them.

The following series of oxides will serve to indicate how gradually the one character disappears as the other is developed. The symbols in black type represent basic oxides, those in roman represent oxides which go to the negative side of the salts into which they enter.

From Non-metallic to Metallic.

SO_3 WO_3 CrO_3 FeO_3
SO_2 WO_2
    Cr_2O_3 Fe_2O_3 Ni_2O_3
                    No salts.
    CrO **FeO** **NiO** **CuO**
    Extremely
    oxidisable.
                        Cu_2O **Ag_2O** **K_2O**
                        Salts unstable.

METALS.—I. THE METALS OF THE ALKALIS. MONADS.

$$\left.\begin{array}{l}\text{Lithium, Li} = 7\\ \text{Sodium, Na} = 23\\ \text{Potassium, K} = 39\cdot 1\end{array}\right\}\frac{Li+K}{2}=23$$

$$\left.\begin{array}{l}\text{Potassium, K} = 39\cdot 1\\ \text{Rubidium, Rb} = 85\cdot 5\\ \text{Cæsium, Cs} = 133\end{array}\right\}\frac{K+Cs}{2}=86$$

Salt Type, M'Cl.

These are soft, white, light, easily fusible, and somewhat volatile metals. They oxidise rapidly in the air, but differ materially from one another in this respect, their affinity for oxygen increasing with the atomic weight. This is shown especially by the spontaneous inflammability of rubidium, and the impossibility of obtaining cæsium by decomposition of its carbonate with charcoal. They all decompose cold water with formation of a soluble hydroxide and evolution of hydrogen gas. But as a consequence of the inferior energy of lithium and sodium, the heat developed by their action upon water is not sufficient to cause the ignition of the escaping hydrogen, whereas the hydrogen disengaged by potassium inflames instantly, and continues to burn with a purple light, due to the accompanying vapour of the metal.

The hydroxides of these metals are all fusible, scarcely decomposed by heat, but volatilising at high temperatures, very soluble in water, and the solutions caustic to the skin, alkaline to litmus, and absorbing carbonic acid from the air. The hydroxides also saponify oils and fats, and the resulting soaps are alkaline and soluble.

The chlorides, sulphides, sulphates, phosphates, and carbonates of the alkali metals are all soluble in water, and their chlorides and sulphates yield perfectly neutral solutions. The sulphates combine with the sulphate of aluminium and the allied metals, generating highly characteristic double salts, called 'alums,' which crystallise in octahedrons. A crystallised lithium-alum has not yet been described, but this is probably owing to its great solubility.

The alkali salts all communicate intense and characteristic colours to the Bunsen flame, and the spectrum (p. 55) of the light so produced exhibits in each case a comparatively small number of bright lines.

As indicated by the atomic weights, the alkali-metals may be divided into two sub-groups. In the one we have potassium, cæsium, and rubidium, which are distinguished by greater chemical activity and inferior solubility of their salts, notably of the platinochlorides and acid tartrates.

$$K_2PtCl_6 \qquad KHC_4H_4O_6$$
$$Rb_2PtCl_6 \qquad RbHC_4H_4O_6$$
$$Cs_2PtCl_6 \qquad CsHC_4H_4O_6$$

Sodium and lithium, though entirely comparable with potassium, are yet distinguished from it by much feebler chemical energies, by the solubility of their platinochlorides, acid tartrates, and alums, and by different minor peculiarities. Sulphate of sodium, for example, crystallises with ten molecules of water, whilst sulphate of potassium is anhydrous. The carbonates also differ, that of sodium forming large efflorescent crystals containing $10H_2O$, the carbonate of potassium occurring in small deliquescent granules, containing usually about one or two molecules of water.

Both sodium and potassium yield unstable peroxides, Na_2O_2 and K_2O_2, and potassium is even capable of passing to a higher stage of oxidation, the product being a tetroxide, K_2O_4. No salts corresponding with these oxides exist.

METALS.—2. METALS OF THE ALKALINE EARTHS. DIADS.

$$\left.\begin{array}{lll}\text{Calcium,} & \text{Ca} = & 40 \\ \text{Strontium,} & \text{Sr} = & 87\cdot6 \\ \text{Barium,} & \text{Ba} = & 137\end{array}\right\} \frac{\text{Ca} + \text{Ba}}{2} = 88\cdot5$$

Salt Type, $M''Cl_2$.

These three metals, obtained by the electrolysis of their fused chlorides, are yellowish, hard, and fusible, and extremely oxidisable. They are, however, less oxidisable than the metals of the alkalies, though they are still capable of decomposing cold water. They also communicate characteristic colours to flame, and yield spectra (p. 55) which are easily recognisable but somewhat more complex than those of the alkalies generally.

The hydrates of the alkaline earths are white substances which are decomposed by heat into water and the anhydrous oxides. They are far less soluble in water than the alkalies, and are proportionately less caustic, alkaline, and attractive of carbonic acid. They saponify fats, but the resulting soaps are generally insoluble in water.

The salts of barium, strontium, and calcium are, as a rule, perfectly neutral.

The chlorides and sulphides are soluble in water, but the sulphates, phosphates, and carbonates are insoluble.

Oxide of barium or baryta exposed to a current of air or oxygen at a heat short of redness absorbs oxygen, and becomes converted into a peroxide, BaO_2. The corresponding peroxides of strontium and calcium are obtained as white precipitates by adding solution of hydric peroxide to lime or strontia water. All three are resolved by ignition into oxygen gas and a residue of the protoxide. They are also soluble in hydrochloric acid, yielding peroxide of hydrogen and the chloride of the metal, *e.g.*—

$$BaO_2 + 2HCl = BaCl_2 + O_2H_2.$$

Hence it is presumable that they have the same constitution as hydric peroxide, the oxygen atoms partly satisfying each other's attractions, according to the following graphic formulæ:—

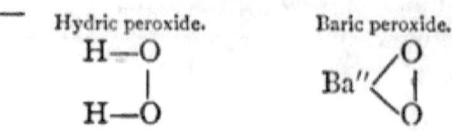

Of the three elements, barium, with the highest atomic weight, is decidedly most basylous. Strontium standing next is, in respect to some characters, more nearly related to barium than to the third member of the series, calcium. Thus the nitrates of strontium and barium crystallise in anhydrous octahedrons, isomorphous with lead nitrate, $Pb(NO_3)_2$. The sulphates of these two metals are also anhydrous, and are practically insoluble in water and acids.

Nitrate of calcium is a deliquescent salt, soluble in alcohol and crystallising in prisms, which contain $Ca(NO_3)_2.4H_2O$. Its sulphate, in the form of gypsum, combines with two molecules of water, is perceptibly soluble in water, and much more freely so in hydrochloric acid.

On the other hand, strontium agrees with calcium in the production of a deliquescent chloride, which is soluble in alcohol, whilst the barium chloride is insoluble in alcohol.

The crystallised chlorides have the following formulæ:—

Chloride of Calcium, $CaCl_2 . 6H_2O$
,, Strontium, $SrCl_2 . 6H_2O$
,, Barium, $BaCl_2 . 2H_2O$

METALS.—3. ZINC GROUP. DIADS.

Magnesium, $Mg = 24$
Zinc, $Zn = 65$
Cadmium, $Cd = 112$

$\dfrac{Mg + Cd}{2} = 68$

Salt Type, $M''Cl_2$.

These elements are rightly associated together in consequence of a very obvious seriation of properties, notwithstanding that they are far less intimately related to one another than, for example, the metals of the alkaline-earth family. Magnesium is a white metal, zinc and cadmium faintly bluish white, and all three are volatile. Their volatility somewhat strangely increases in proportion as the atomic weight increases, whilst their basigenic power diminishes. That zinc is decidedly more positive than cadmium is shown by its power of precipitating cadmium in the metallic state from its solutions. That magnesium is more positive than the other two is shown by its power of decomposing water when heated with it, or more readily if previously coated with pulverulent copper.[1] Also by the precipitation of both zinc and cadmium when metallic magnesium is introduced into solutions of their salts.

Magnesium, zinc, and cadmium are all easily combustible in air or oxygen, the combustion of the former two being attended by the emission of a dazzling light. Each metal forms one oxide : MgO, a white unalterable powder ; ZnO, a white powder, becoming yellow when heated ; CdO, a yellowish brown powder. These oxides are insoluble in water, that of magnesium only showing a faint alkaline reaction when placed upon wet test-paper. The oxides and hydroxides are readily soluble in solutions of ammoniacal salts.

The hydroxides are easily resolved, by being heated, into water and the oxide, and the carbonates in like manner give up carbonic anhydride, leaving a residue of the oxide.

The chlorides are volatile, deliquescent solids.

The sulphide of magnesium is an earthy substance which is decomposed even by water—

$$MgS + 2OH_2 = Mg(OH)_2 + SH_2$$

and is therefore not precipitated on the addition of a soluble sulphide to a magnesian solution.

[1] Zinc and cadmium, when coated with spongy copper, are also capable of decomposing water slowly.

Zinc sulphide (native = blende) is a white precipitate easily soluble in diluted mineral acids, and hence only imperfectly precipitated by the action of hydrogen sulphide upon the solution of a zinc salt.

Cadmium sulphide (native = greenockite) is a yellow precipitate thrown down by hydrogen sulphide from acidified solutions of cadmium salts.

The sulphates of these metals are, perhaps, the most characteristic of their salts. They are soluble in water, and have the following formulæ :—

$$\begin{array}{ll} \text{Magnesium sulphate} & MgSO_4.7H_2O \\ \text{Zinc sulphate} & ZnSO_4.7H_2O \\ \text{Cadmium sulphate} & CdSO_4.4H_2O \end{array}$$

The magnesium and zinc salts crystallise in four-sided prisms, isomorphous with the corresponding nickel sulphate.

All three combine with potassium sulphate, generating double salts, which crystallise with six molecules of water:

$$MgSO_4 \cdot K_2SO_4 \cdot 6H_2O$$
$$ZnSO_4 \cdot K_2SO_4 \cdot 6H_2O$$
$$CdSO_4 \cdot K_2SO_4 \cdot 6H_2O$$

Sulphates isomorphous with these are produced by several other metals, such as copper and iron, and will be referred to in the proper place.

A general review of its properties indicates that magnesium forms a connecting link between zinc and calcium. From the latter it differs in the insolubility of its hydroxide in water, and in the solubility of the same compound in ammonium chloride. Sulphate of magnesium is also distinguished from sulphate of calcium by its ready solubility in water.

METALS.—4.

Silver, Ag = 108. Mercury, Hg = 200.
Salt type, M'Cl, *Salt types, M''_2Cl_2,*
or M''_2Cl_2? *and $M''Cl_2$.*

Silver is a white metal, very malleable and ductile, fusible at a red heat, and volatilising in blue vapour at very high temperatures. It is the best known conductor of heat and electricity. It is undoubtedly related to the metals of the alkalies, for its chloride is isomorphous with sodium chloride, its sulphate is isomorphous with anhydrous sodium sulphate, and by combination with aluminium sulphate it yields a true alum, crystallising in octahedrons. In almost every other respect, however, it differs from them.

Thus its relative density is high, 10·5, and it is quite untarnished by pure air, whether dry or moist, the blackening so often observed being due to the formation of a film of sulphide. Pure silver has the singular property of occluding, when fused in the air, a considerable volume of oxygen, which escapes as the temperature goes down and the metal solidifies. Silver has no action upon water or steam at any temperature, and is quickly reduced from its salts by zinc, iron, or even mercury. The oxide is said to be sufficiently soluble in water to give a faintly alkaline reaction, but it differs entirely from potash or soda in appearance and in its decomposability by heat. The chloride, iodide, sulphide, carbonate, and phosphate of silver are all insoluble in water, and the sulphate very sparingly soluble.

Silver forms a peroxide, Ag_2O_2, even less stable than the potassium peroxide. There is also a suboxide, Ag_4O, the composition of which is, however, not very well established.

The relations of silver to the alkali-metals on the one hand, and to mercury on the other, are about equally distant. This connection is obscurely indicated by comparison of their atomic weights, from which it may be seen that silver is intermediate between sodium and mercury.

$$\frac{\text{Na } 23 + \text{Hg } 200}{2} = 111\cdot5 \ (\text{Ag} = 108).$$

Mercury, like silver, is a white metal, crystallising in the regular system and volatilising, though much more readily. Both are easily reducible to the metallic state, and the metals are both incapable of decomposing water or hydrochloric acid. Mercury, however, is slowly oxidised by heating in air or oxygen. Both are soluble in nitric acid, and when boiled with concentrated sulphuric acid sulphur dioxide is in each case evolved, and a sulphate produced in the form of a sparingly soluble crystalline powder.

Silver oxide, Ag_2O, and mercurous oxide, Hg_2O, are both nearly black, insoluble in water, and resolved by heat into oxygen gas and the metal.

Silver chloride, $AgCl$, and mercurous chloride, Hg_2Cl_2, are both obtainable by precipitation as white powders, which blacken, the silver salt quickly, the mercury salt slowly, on exposure to light.

The iodides are both yellow and insoluble. The chief distinction between silver and mercurous salts lies in the volatility of the latter.

As regards the formulæ of the mercurous compounds, their analogy to the compounds of silver points to the adoption of similar formulæ.

$$\text{Ag} - \text{O} - \text{Ag}. \qquad \text{Hg} - \text{O} - \text{Hg}.$$
$$\text{Ag} - \text{Cl}, \ \&c. \qquad \text{Hg} - \text{Cl}, \ \&c.$$

But, inasmuch as the low vapour density of calomel, $117\cdot75 = \frac{200 + 35\cdot5}{2}$, has been traced to dissociation (Chap. XVIII.), this, and consequently all other mercurous compounds, must be represented as containing diad mercury, thus:

$$\begin{array}{c}\text{Hg}\\|\\\text{Hg}\end{array}\!\!>\!\text{O} \qquad \begin{array}{c}\text{Hg}-\text{Cl}\\|\\\text{Hg}-\text{Cl}\end{array} \quad \&c.$$

The relationship subsisting between them and the corresponding salts of silver is best recalled by using similar formulæ for the latter, thus:

$$\begin{array}{c}\text{Ag}\\|\\\text{Ag}\end{array}\!\!>\!\!\text{O} \qquad \begin{array}{c}\text{Ag—Cl}\\|\\\text{Ag—Cl}\end{array} \qquad \&\text{c}.$$

Such a view is also consistent with the observed equivalency of silver and copper in the sulphides, Ag_2S, Cu_2S, and $CuAgS$. On the other hand, the connection already referred to between silver and monad sodium must not be overlooked.

The mercuric salts are almost without parallel as regards properties among metallic salts. Their formulæ are written upon the same type as those of the salts of cadmium and zinc, but it cannot be said that the resemblance extends much farther.

In one respect, however, metallic mercury agrees with the metals cadmium and zinc. Notwithstanding its much greater volatility, its vapour density, like that of the other two, is the half of its atomic weight. Hence, whilst the molecules of all other volatile elements are polyatomic, those of the three metals referred to are simply monatomic, the atom and the molecule being represented in each case by one and the same symbol or formula—Zn, Cd, Hg.

The following are the most important mercuric compounds. All are volatilisable.

Oxide, HgO. A red or yellow powder darkening by heat, and decomposed below redness into Hg and O_2.

Chloride, $HgCl_2$. White, crystallisable, soluble in water, alcohol, and ether. Vapour density $\frac{200 + 71}{2} = 135\cdot5$.

Amidochloride, $HgNH_2Cl$. Commonly known as 'White Precipitate.' Representable either as an amidochloride,

$$\text{Hg}''\!\!<\!\!\begin{array}{c}\overset{\prime\prime\prime}{\text{N}}=\text{H}_2\\\text{Cl}\end{array}$$

or as chloride of mercuric ammonium,

$$\overset{v}{N} \overset{\displaystyle Hg}{\underset{\displaystyle Cl}{=\!=\!= H_2}}$$

The former view is preferable, considering its formation from mercuric chloride by ammonia.

$$Hg{<}^{Cl}_{Cl} + {}^{NH_2H}_{NH_3} = Hg{<}^{NH_2}_{Cl} + \{{}^{HCl}_{NH_3}$$

Iodide, HgI_2. Red or yellow, dimorphous, insoluble in water, soluble in ether and in solution of potassic iodide.

Sulphide, HgS. Cinnabar, dark red masses of hexagonal prisms. Artificial vermilion, bright red powder.

Sulphate, $HgSO_4$. Colourless, minute, prismatic crystals, characteristically decomposed by water with formation of yellow 'turpeth mineral,' or trimercuric orthosulphate, $Hg_3O_2SO_4$ or $Hg''_3(SO_6)^{vi}$.

METALS.—5.

Thallium, $Tl = 204$.
Triad. Salt types $M'Cl$, and $M'''Cl_3$.

Lead, $Pb = 207$.
Tetrad. Salt type, $M''Cl_2$.

THALLIUM was originally discovered and occurs most abundantly in certain kinds of copper and iron pyrites. It is a very soft, nearly white metal, which tarnishes quickly but superficially in the air. It closely resembles lead in physical properties, except that it is softer and streaks paper more readily. It melts at 294°, and volatilises at a red heat. It is capable of burning brilliantly in oxygen.

Thallium forms two classes of compounds, those in which it is univalent being the more stable.

The following are the most characteristic of *thallous* compounds :—

Chloride, TlCl. A white curdy precipitate, resembling silver chloride, but crystallisable from boiling water, and combining with platinic chloride to form a yellow, slightly soluble, crystalline compound, Tl_2PtCl_6.

Oxide, Tl_2O, and *Hydroxide*, TlHO. These compounds are soluble in water, the solution being strongly alkaline, and reacting in the same manner as caustic potash. The hydroxide differs from the alkali hydroxides in losing the elements of water when exposed over oil of vitriol in a vacuum. The residual oxide is almost black, and after fusion crystalline.

Sulphate, Tl_2SO_4. A soluble colourless salt, isomorphous with sulphate of potassium. It combines with aluminium sulphate, forming a true alum, $TlAl(SO_4)_2.12H_2O$.

Sulphide, Tl_2S. A brown precipitate, soluble in acids.

Thallic Compounds.—*Chloride*, $TlCl_3$. A soluble salt, crystallisable by evaporation of its solution in a vacuum. It melts and evolves chlorine at high temperatures. It combines with the monochloride, forming two compounds, Tl_4Cl_6 or $3TlCl.TlCl_3$, and Tl_2Cl_4 or $TlCl.TlCl_3$. It forms similar compounds with the chlorides of potassium and ammonium.

Oxide, Tl_2O_3. A dark red powder, insoluble in water, and reduced by heat to thallous oxide. Thallium is readily obtained from its salts either by electrolysis, by the action of zinc, or by fusion with cyanide of potassium.

The ordinary spectrum of thallium is the simplest known. It consists of a single bright band in the green.

LEAD is a bluish metal, so soft as to streak paper, and possessing very little tenacity. Its specific gravity is 11·36, a number very near to, and intermediate between, that of thallium, 11·9, and of silver, 10·53. The latter metal is very constantly associated with it, occurring as sulphide in, probably all, galena. Lead melts at 326°, and at a red heat volatilises freely. A freshly cut surface tarnishes rapidly in air, but oxidation, except of the melted metal, does not proceed to any appreciable extent.

Lead forms two oxides of definite composition, namely, litharge, PbO, and the peroxide, PbO_2, besides several mixed oxides containing the elements of these two in various proportions. The red oxide is generally $2PbO, PbO_2$, or Pb_3O_4. When plumbic peroxide is acted upon by hydrochloric acid, it does not, like barium peroxide, give rise to hydrogen peroxide, but to a tetrachloride, $PbCl_4$. In these two compounds, therefore, lead is quadrivalent, and their formulæ must be written as follows:—

$$O = Pb = O \qquad \begin{array}{c} Cl \\ Cl \end{array}\!\!>\!Pb\!<\!\!\begin{array}{c} Cl \\ Cl \end{array}$$

The tetrachloride is a very unstable compound, and no oxy-salts corresponding with it are known.

The ordinary salts of lead are typified by the *dichloride*, $PbCl_2$, a white, sparingly soluble salt, crystallisable from boiling water.

The *oxide*, PbO, is a dull yellow, the hydroxide a white powder. Both are slightly soluble in water, the solubility, like that of lime, being increased by the addition of sugar. The solutions are strongly alkaline, and absorb carbonic acid from the air.

Sulphate, $PbSO_4$. In the form of lead-vitriol, isomorphous with heavy spar, $BaSO_4$.

Carbonate, $PbCO_3$. As white-lead ore or cerusite, isomorphous with witherite, $BaCO_3$.

Nitrate, $Pb(NO_3)_2$, crystallises in octahedrons, isomorphous with barium nitrate.

Sulphide, PbS. Black insoluble powder or crystallised in shining cubes (galena).

Relations of Thallium and Lead.

The relations of thallium both to potassium and lead are exceedingly well-marked. With potassium it agrees in the alkalinity of its oxide and the characters of the platino-chloride, acid tartrate, sulphate, and alum. But it differs from potassium and resembles lead in the ready reducibility of the metal, in its high atomic weight and density, and in the characters of the sulphide, monochloride, and other salts.

Lead, in its turn, has strong points of resemblance on the one hand to thallium, and on the other to barium, as indicated by the isomorphism of many lead and barium salts.

The relations which are thus manifested by these metals serve to bring into view the connection which they undoubtedly have with other metals, the characters of which have already been discussed. Without attempting to trace out any further parallels of the same kind, a task which the student can very well perform without assistance, we may just indicate the direction in which analogies will be most readily detected by arranging the symbols of the most important of these metals in the following order :—

$$
\begin{array}{cccc}
(Hg_2)'' \text{——} Hg'' & Pb'' \text{——} Tl' \\
Ag & Cd & Ba & Cs \\
| & Zn & Sr & Rb \\
Na & Mg - Ca & K \\
\end{array}
$$

METALS.—6. ALUMINIUM GROUP.

Aluminium, Al = 27
Gallium, Ga = 69
Indium, In = 114

Triads or quasi-tetrads. Salt type, MCl_3 or $(M_2)^{vi}Cl_6$.

These are white metals of comparatively low density, standing towards one another in a position which is the counterpart of that shown by the group magnesium, zinc, cadmium.

Aluminium is a very abundant constituent of various silicates. It forms a single oxide and a series of salts which are closely related to the corresponding compounds of iron, as will be shown in the next group.

Gallium and indium, on the other hand, are very rare elements occurring in certain zinc ores. They differ from aluminium in their lower fusing points and in their more pronounced triad character, also in the formation of more stable sulphides. The following tabular statement will recall their chief characteristics.

	Aluminium	Gallium	Indium
Density	2·6	5·9	7·4
Melting point	750°–800°	30°	176°
Chloride	$AlCl_3$ (at 760° and upwards)	$GaCl_3$	$InCl_3$
Oxide	Al_2O_3 white	Ga_2O_3 white	In_2O_3 yellow
Alum $\left.\begin{smallmatrix}M'\\M'''\end{smallmatrix}\right\}(SO_4)_2\,12H_2O$	Octahedral	Octahedral	Octahedral
Sulphide	Decomposed by water	Precip. only in company with ZnS	Yellow precipitate

Metals related to Iron.

METALS.—7. IRON-COPPER GROUP.

SUB-GROUP A.

Aluminium, Al = 27.
Tetrad. Salt type $(M_2)^{VI}Cl_6$.

Chromium, Cr = 52·5.
Hexad. Salt types, $M''Cl_2$ and $(M_2)^{VI}Cl_6$.

Manganese, Mn = 55 }
Iron,   Fe = 56 }
Hexad. Salt types, $M''Cl_2$ and $(M_2)^{VI}Cl_6$.

Cobalt, Co = 58·7 }
Nickel, Ni = 58·7 }
Tetrad. Salt type, $M''Cl_2$.

These are white or grey metals of very high melting point. Their relative densities follow one another in the same order as the atomic weights, and are almost directly proportional to them. Including copper, the solitary member of the next sub-group, they may be placed in the following order, which we shall show presently represents very fairly their mutual chemical relations :

Atomic Weights.		Relative Densities.
27	Al	2·6
52·1	Cr	6·8 to 7·3

Atc. Wt.		Rel. Dens.	Atc. Wt.		Rel. Dens.
54	Mn	8·0 (about)	56	Fe	7·8
59	Co	8·9	58	Ni	8·6
			63·3	Cu	8·9

T

Al	Cr	Fe
No Aluminous Compounds known.	$CrCl_2$ $CrSO_4 7H_2O$? pale blue $\left.\begin{array}{l}CrSO_4\\K_2SO_4\end{array}\right\}$ $6H_2O$	FeO $FeCl_2$ $FeSO_4 \cdot 7H_2O$ pale green $\left.\begin{array}{l}FeSO_4\\K_2SO_4\end{array}\right\} \cdot 6H_2O$
		isomorphous with
All Aluminic Compounds Colourless.	Cr_3O_4	Fe_3O_4
Al_2O_3	Cr_2O_3	Fe_2O_3
Al_2Cl_6 volatile.	Cr_2Cl_6 violet: volatile at red heat.	Fe_2Cl_6 green: volatile at red heat
$Al_2(SO_4)_3$	$Cr_2(SO_4)_3$	$Fe_2(SO_4)_3$
$AlK(SO_4)_2 \cdot 12H_2O$	$CrK(SO_4)_2 \cdot 12H_2O$	$FeK(SO_4)_2 \cdot 12H_2O$
		modified — (FeS_2)
	CrO_2Cl_2	—
	CrO_3	—
	CrO_4K_2	FeO_4K_2
	CrO_4Ba	FeO_4Ba
	$\left\{\begin{array}{l}CrO_4K_2\\CrO_3\end{array}\right.$ or $\left\{\begin{array}{l}CrO_2 \cdot OK\\O\\CrO_2 \cdot OK\end{array}\right.$	—
	$\left\{\begin{array}{l}CrO_4K_2\\2CrO_3\end{array}\right.$ or $\left\{\begin{array}{l}CrO_2OK\\O\\CrO_2\\O\\CrO_2OK\end{array}\right.$	—
	$Cr_2O_6(OH)_2$?	—

Metals related to Iron.

Ni	Mn	Co
NiO NiCl$_2$ NiSO$_4$.7H$_2$O bright green	MnO MnCl$_2$ MnSO$_4$.7H$_2$O pale pink	CoO CoCl$_2$ CoSO$_4$.7H$_2$O bright red
$\left.\begin{array}{l}\text{NiSO}_4\\ \text{K}_2\text{SO}_4\end{array}\right\}$·6H$_2$O	$\left.\begin{array}{l}\text{MnSO}_4\\ \text{K}_2\text{SO}_4\end{array}\right\}$6H$_2$O	$\left.\begin{array}{l}\text{CoSO}_4\\ \text{K}_2\text{SO}_4\end{array}\right\}$6H$_2$O

corresponding magnesium salts.

Ni	Mn	Co
—	Mn$_3$O$_4$	Co$_3$O$_4$
Ni$_2$O$_3$	Mn$_2$O$_3$	Co$_2$O$_3$
No Salts	Mn$_2$Cl$_6$ decomposed by heat	Co$_2$Cl$_6$ decomposed by heat
—	Mn$_2$(SO$_4$)$_3$?	Co$_2$(SO$_4$)$_3$?
—	MnK(SO$_4$)$_2$12H$_2$O	—

cubes or regular octahedrons.

Ni	Mn	Co
—(NiS$_2$)	MnO$_2$	—
—	MnO$_2$Cl$_2$	—
—	MnO$_3$	—
—	MnO$_4$K$_2$	—
—	MnO$_4$Ba	—
—	—	—
—	—	—
—	—	—
—	—	—
—	Mn$_2$O$_3$(OH)$_2$ and Mn$_2$O$_7$	—

The metals of sub-group A resist atmospheric oxidation to a considerable extent, except moisture or carbonic acid be present. Several of them, especially iron and manganese, are capable of decomposing steam at a red heat with evolution of hydrogen. They all dissolve more or less rapidly in dilute acids.

Iron, cobalt, and nickel are strongly paramagnetic, chromium and manganese feebly so.

A synoptical view of the chemical relations of these metals to one another will be best obtained, since they are somewhat complex, by tabulating the formulæ of their most characteristic compounds as on the preceding page. Few observations upon the table are necessary. As in all similar groups of intimately related elements, each one seems to affect a particular state of oxidation or combination in which it attains a condition of chemical repose or equilibrium. Thus the aluminic salts are absolutely irreducible to any lower state of oxidation; chromic salts reducible with great difficulty; ferric salts are easily transformed into ferrous, whilst manganic salts evolve chlorine or oxygen by mere ebullition. The isomorphous relations of these metals to one another are of great interest, and are exhibited not only by the alums and other artificial salts, but in the constant association of these metals in nature, as, for example, in the replacement of aluminium by iron in clays, and in the silicates from which they are formed, in the replacement of the elements of ferric oxide by chromic oxide in chrome-iron, $FeO.Cr_2O_3$, and by the occurrence of metallic nickel in meteoric iron.

Several members of the group exhibit peculiarities which are well worthy of study, but in this place can receive only passing notice. Such, for example, are the metameric (?) modifications of alumina and of chromium salts, indicated in the latter case by curious changes of colour and solubility; also the remarkable influence of a small quantity of carbon in increasing the fusibility and hardness of iron, nickel, cobalt, and, perhaps, others of the group; also the production of an extensive series of ammoniacal bases by cobalt, not by nickel.

Valency of the Iron Group.

Manganese forms a sexfluoride, MnF_6, and a class of salts, the manganates, which are said to be isomorphous with the chromates and sulphates. Hence manganese is sexivalent, and the formulæ of manganic perfluoride and potassic manganate may be written as follows:

Manganese also forms a dioxide in which it is quadrivalent,

$$O = Mn = O,$$

and a chloride corresponding with ferrous chloride in which it may be considered to be quadrivalent,

$$Cl_2 = Mn = Mn = Cl_2$$

or bivalent, $\quad Cl - Mn - Cl.$

There is no direct evidence for either the one formula or the other, and, consequently, the latter is very commonly accepted for the sake of simplicity.

The permanganates must be represented by the formula $M'_2Mn_2O_8$, on the assumption of the hexad character of manganese, so that potassic permanganate becomes—

$$\begin{array}{c} O \\ \parallel \\ O=Mn-O-OK \\ | \\ O=Mn-O-OK \\ \parallel \\ O \end{array}$$

or more probably,

$$\begin{array}{ccc} \text{OK} & & \text{OK} \\ | & & | \\ \text{O}=\text{Mn}-\text{O}-\text{O}-\text{Mn}=\text{O} \\ \| & & \| \\ \text{O} & & \text{O} \end{array}$$

The alleged isomorphism of the permanganates and perchlorates is explainable on the assumption either that chlorine is hexad like manganese and perchloric acid is expressed by the formula $H_2Cl_2O_8$, or that both chlorine and manganese are triad or heptad in the formulæ $HClO_4$ and $HMnO_4$, thus

$$\text{H}-\text{O}-\text{Cl}\begin{array}{c}\diagup\text{O}\diagdown\\\diagdown\text{O}\diagup\end{array}\text{O} \quad \text{or} \quad \text{H}-\text{O}-\overset{\overset{\text{O}}{\|}}{\underset{\underset{\text{O}}{\|}}{\text{Cl}}}=\text{O}$$

By reason of the correspondence between the ferrates and manganates, iron must also be regarded as hexad; although in ferric, and probably also ferrous compounds, it is usually assumed to be tetrad.

Thus the vapour density of ferric chloride at moderate temperatures [1] agrees with the half of the molecular weight denoted by the formula Fe_2Cl_6, although the formula $FeCl_3$ would equally well record its composition. The existence of iron pyrites, FeS_2, the analogue of SnS_2, also supports the hypothesis of the quadrivalence of iron in these compounds.

$$\begin{array}{cc}\text{Cl} & \text{Cl}\\ | & |\\ \text{Cl}-\text{Fe}-\text{Fe}-\text{Cl}\\ | & |\\ \text{Cl} & \text{Cl}\end{array} \qquad \text{S}=\text{Fe}=\text{S}$$

Chromium is also tetrad in its ordinary chromic salts, but

---

[1] At high temperatures decomposition takes place into ferrous chloride and chlorine.

hexad in the chromates, as well as in chromium sexfluoride, trioxide, and oxychloride.

$$
\begin{array}{cc}
\text{Cl} \quad \text{Cl} \\
| \quad | \\
\text{Cl}-\text{Cr}-\text{Cr}-\text{Cl} \\
| \quad | \\
\text{Cl} \quad \text{Cl}
\end{array}
\qquad
\begin{array}{c}
\text{KO} \diagdown \quad \diagup \!\!\!\! / \text{O} \\
\text{Cr} \\
\text{KO} \diagup \quad \diagdown \text{O}
\end{array}
$$

Chromic chloride.   Potassium chromate.

The ordinary salts of cobalt and nickel correspond to the ferrous and zinc salts, but they form unstable peroxides, Co_2O_3 and Ni_2O_3, which lose oxygen at a red heat.

These compounds, and the salts of the oxycobaltamines are almost the only known tetradic combinations of these two metals.

Sub-group B.

Copper, $\dot{C}u = 63\cdot 3$.

Diad. Salt-types, $(M_2)''Cl_2$ and $M''Cl_2$.

This metal is distinguished from the members of the series just considered by its red colour, by its very superior conductivity of heat and electricity, by its diamagnetic properties, by its ready oxidation when heated in dry air, and by the easy reduction of the metal from its oxide and salts, also by its indifference to diluted sulphuric or hydrochloric acid. It is further characterised by the fact that no oxide superior to the oxide CuO can be isolated, by the production of a suboxide, Cu_2O, and corresponding series of salts, also by the facility with which all its compounds unite with or act upon ammonia. By the last two characters it indicates its relationship to silver and mercury.

The following are the most important of the compounds of copper, together with the formulæ of compounds with which they manifest greater or less analogy :—

Cuprous.	Mercurous.	Argentic.
Cu_2O	Hg_2O	Ag_2O
or	or	
$Cu\!\!-\!\!\overset{\displaystyle \mid}{}\!\!\searrow O$ $Cu\!\!\nearrow$	$Hg\!\!-\!\!\overset{\displaystyle \mid}{}\!\!\searrow O$ $Hg\!\!\nearrow$	$Ag\!\!-\!\!\overset{\displaystyle \mid}{}\!\!\searrow O$ $Ag\!\!\nearrow$
Cu—Cl | Cu—Cl	Hg—Cl | Hg—Cl	Ag—Cl | Ag—Cl
Cu—H | Cu—H	No hydride.	No hydride.

Cu_2S isomorphous with Ag_2S and with $CuAgS$.

Very few cuprous oxysalts are known, for when cuprous oxide is acted upon by acids it generally splits up and yields a cupric salt and a residue of metallic copper.

Cupric.	Iron and Manganese.
CuO	FeO
$CuSO_4.5H_2O$ isomorphous with	$FeSO_4.5H_2O$
and with	$MnSO_4.5H_2O$
$\left.\begin{array}{l}CuSO_4\\K_2SO_4\end{array}\right\}6H_2O$ isomorphous with	$\left.\begin{array}{l}FeSO_4\\K_2SO_4\end{array}\right\}.6H_2O$
CuS	FeS
$FeCuS_2$	
or $Fe_2Cu_2S_4$	FeS_2
Copper pyrites.	Iron pyrites.

Considering the very general blue or green colour of the cupric salts, and their dissolution in excess of ammonia to form a blue liquid, copper approaches more nearly to nickel than to any other member of the iron group. And although a pentahydrated nickel sulphate corresponding with common blue vitriol has not been described, yet of the isomorphism of copper and nickel in the form of sulphate there can be no doubt. For when a mixture of cupric and nickel sulphates is crystallised, the crystals which are deposited, though containing both nickel and copper, have the form of nickel sulphate, and contain seven molecules of water when the nickel is in excess. If, on the other hand, the cupric sulphate is in excess, the crystals have the form of blue vitriol and contain five molecules of water.

Through the isomorphism of the double potassio-sulphates, the iron-copper group of metals is connected with the zinc group previously described.

METALS.—8. PLATINUM GROUP.

SUB-GROUP A.—GOLD.

Triad. Salt types ($\overset{\prime\prime}{M_2}$)$Cl_2$ *and* $\overset{\prime\prime\prime}{M}Cl_3$.

Gold is a metal of familiar yellow colour and high density. In consideration of its inalterability in the fire or by the action of acids (save nitro-muriatic acid, hence called *aqua regia*), gold was regarded by the old chemists as the type of a 'noble' metal. It forms two classes of salts, of which the aurous are somewhat unstable, readily undergoing decomposition into an auric compound and metallic gold. The sodio-aurous thiosulphate $\left.\begin{matrix}Na_6\\Au_2\end{matrix}\right\}(S_2O_3)_4.4H_2O$ is one of the most definite.

The trichloride is the most important auric compound. It is a red, crystalline, deliquescent body, which forms yellow crystallisable compounds with the chlorides of hydrogen and the alkali-metals, $HAuCl_4$, $KAuCl_4$, NH_4AuCl_4, &c.

Gold is reduced from its solutions with the greatest readiness by nitrous acid, sulphurous acid, ferrous salts, and all reducing agents, including many organic substances. Its oxides and chlorides also yield up the metal when strongly heated.

Sub-group B.

The members of this group range themselves very naturally into two divisions, each of which exhibits a striking uniformity of atomic weight and density. With one of these gold allies itself, in virtue of its high density and atomic weight, as well as its resistance to oxidation and solution in acids; also by the yellow colour of its compounds and the tendency of its highest chloride to form double salts. With the other, silver is, perhaps, in a similar manner, remotely connected, though in this case it must be admitted there is but little in the properties of the metal to sanction such an arrangement.

	Symb.	At. Wt.	Dens.		Symb.	At. Wt.	Dens.
Silver *Monad*	Ag	108	10·5	Gold *Triad*	Au	197	19·3
Palladium	Pd	106	11·8	Platinum	Pt.	195	21·5
Rhodium	Ro	104	11·0	Iridium	Ir	193	21·1
Ruthenium *Tetrad*	Ru	104	11·4	Osmium *Tetrad*	Os	199	21·4

The true platinum metals are white metals which require for their fusion the highest attainable temperatures of the oxyhydrogen flame. It is stated, indeed, that osmium has never been really melted; hence the numbers by which their respective densities are denoted can only be approximately correct.

They differ somewhat in their behaviour towards reagents. Palladium, for example, is slowly dissolved by nitric acid, whilst some of the others resist the action even of *aqua regia*. Ruthenium and osmium, on the other hand, although slowly attacked by *aqua regia* when in the compact state, undergo oxidation somewhat rapidly when heated in the air.

They all, with the exception of rhodium, form more or less stable tetrachlorides, which are generally yellow and easily reducible by heat, either to the metallic state or to lower chlorides. These compounds also unite with the chlorides of the alkali-metals, forming crystallisable salts, which in some cases, as, for example, the platinochlorides of potassium and ammonium,

$$2KCl.PtCl_4 \text{ or } K_2PtCl_6$$
$$2NH_4Cl.PtCl_4 \text{ or } (NH_4)_2 PtCl_6$$

are almost insoluble and highly characteristic. But beside these tetrachlorides, the platinum metals yield dichlorides and sesquichlorides, which are in special cases of greater importance. The dichloride of palladium, for example, is one of its most important compounds, and far more stable than the tetrachloride, which is known only in a state of solution, and in its double salts. Rhodium, again, forms only one chloride, and that the sesquichloride.

The following are all the definite chlorides of these metals which have been described :—

Dichlorides.	Sesquichlorides.	Tetrachlorides.
$PdCl_2$	—	$PdCl_4$
$PtCl_2$	—	$PtCl_4$
—	Ro_2Cl_6	—
$IrCl_2$?	Ir_2Cl_6	$IrCl_4$
$RuCl_2$	Ru_2Cl_6	$RuCl_4$
$OsCl_2$	Os_2Cl_6	$OsCl_4$

Ruthenium and osmium are specialised by the existence of potassium salts called rutheniate and osmiate, which contain the elements of potassium oxide, together with a trioxide of ruthenium or osmium.

$$K_2O.RuO_3 \text{ or } K_2RuO_4.$$
$$K_2O.OsO_3 \text{ or } K_2OsO_4.$$

The trioxides, or anhydrides, are not known in an isolated state. But the most remarkable compounds of these metals are the tetroxides RuO_4, OsO_4, which are highly volatile bodies, soluble in water to form acid solutions, but apparently incapable of generating salts. The constitution of these compounds is not well understood. They are the only tetroxides known.

CHAPTER XXVI.

THE PERIODIC LAW.

THROUGHOUT the preceding chapters, in which the classification of the elements and of their compounds has been discussed, frequent reference has been made to the remarkable relations which are observable in the numerical values of the atomic weights. These relations have from the first offered to chemists an attractive problem, the importance of which has greatly increased within the last few years.

Very early in the history of modern chemistry, namely, in the year 1815, the hypothesis was put forward by Dr. Prout that, taking the atomic weight of hydrogen as 1, the atomic weights of all the remaining elements are multiples of this by a whole number. This hypothesis, either in its original form or modified subsequently by the assumption of the value ·5 or ·25 for the atomic weight of hydrogen, has not stood the test of rigorous experimental investigation. But within the last few years a further modification of Prout's hypothesis has attracted attention in connection with speculations relating to the genesis of the so-called elements, based upon the fundamental idea that these bodies were formed in the beginning by the condensation or polymerisation of the primordial simple matter which is supposed to have pervaded all space.

Independently of such considerations, however, the relations which have been discovered to subsist between the atomic weights of the elements and the properties of these substances and of their compounds have been made the basis of a system of classification which has been almost universally adopted by chemists. A short account of some of the most interesting of these observations will now be laid before the student.

A relation which is conspicuous amongst the atomic weights of the best known elements is one which has already been frequently adverted to in previous pages. It is this. Many closely related elements may be ranged in short series of three, in which a gradual modification of properties is observable in tracing the series through its successive terms. In such series the atomic weight of the intermediate term is generally either the half of the sum of the atomic weights of the extremes, or is very near to such number. In the halogens, for example, the atomic weight of bromine is 80, whilst the atomic weight of chlorine is 35·5 and that of iodine 127. The mean of these two latter numbers is 81·25.

A still closer agreement with the mean of the extremes in such a series is shown in the case of the alkali metals, lithium, sodium, and potassium.

$$\text{Atomic weight of lithium} = 7$$
$$\text{,, ,, sodium} = 23$$
$$\text{,, ,, potassium} = 39·1$$

And $\frac{7 + 39·1}{2} = 23·05.$

In a few cases the atomic weights of elements which are associated together by reason of their chemical characteristics are simple multiples one of another, as in the example of oxygen, whose atomic weight is approximately 16, whilst that of sulphur is 32, or they have nearly the same value, as in the case of iron and manganese, nickel and cobalt.

These considerations have been very ably discussed by several chemists, but especially by Dumas, who has shown that one of the most interesting relations among the atomic weights of the elements may be expressed somewhat as follows. The numbers representing the atomic weights of a series of closely allied elements, when written down in the order of their numerical value, represent an arithmetical progression which may be expressed by a common formula,

$$a + nd,$$

in which a is the first term and d the difference in passing from term to term. Thus, for example, the series of alkali metals conforms to this type:

$$a = 7 \qquad d = 16$$

Atomic weight of lithium $= a = 7$
,,   ,,   sodium $= a + d = 23$
,,   ,,   potassium $= a + 2d = 39$

Now it is obvious that this relationship is just that which obtains in the combining values of the several members of a homologous series. The radicles of the series to which common alcohol belongs, for example, run in the following order, where $a = 15$ and $d = 14$:

Methyl $CH_3 = 15 = a$
Ethyl $C_2H_5 = 29 = a + d$
Propyl $C_3H_7 = 43 = a + 2d$
Butyl $C_4H_9 = 57 = a + 3d$
Amyl $C_5H_{11} = 71 = a + 4d$
&c.   &c.   &c.   &c.

Such a series also exhibits several of the peculiarities already pointed out in several families of elements. Thus, taking three contiguous terms or three terms at equal intervals in the series, the middle one has a combining weight equal to half the sum of the combining weights of the two extreme terms. Thus:

$$\frac{CH_3 + C_3H_7}{2} = \frac{15 + 43}{2} = 29 = C_2H_5.$$

Moreover in the higher members of such a series numbers such as 141 and 281, 127 and 253, 113 and 225 occur, and it may be argued that if these radicles were elementary or if their composition were unknown, such numbers might fairly be considered as really representing a ratio of 1 : 2.

This simple formula, however, does not account for more than a few of the elementary series, and in many cases it is necessary to recognise a somewhat more complex expression, which may be written

$$a + nd + nd' + nd'' + \ldots$$

in which a represents the fundamental characteristic of the series, and d, d', d'', d''', &c., the differences.

An example or two will render its application easily intelligible.

THE HALOGENS.

General Formula.

First term a
Second „ $a + d$
Third „ $a + 2d + d'$
Fourth „ $a + 2d + 2d' + d''$

NUMERICAL VALUES.

First term 19 $= F$
Second „ $19 + 16·5$ $= 35·5 = Cl$
Third „ $19 + 33 + 28$ $= 80 = Br$
Fourth „ $19 + 33 + 56 + 19 = 127$ $= I$

Here $d'' = a$, and the fourth term therefore becomes

$$2a + 2d + 2d'.$$

A similar general formula will be found to apply to the series N, P, As, Sb, Bi. Here $a = 14$, $d = 17$, $d' = 27$, $d'' = 18$, $d''' = 2a$.

Nitrogen $= a$ $= 14$
Phosphorus $= a + d$ $= 31$
Arsenic $= a + 2d + d'$ $= 75$
Antimony $= a + 2d + 2d' + d''$ $= 120$
Bismuth $= a + 2d + 2d' + 4d'' + d'''$
or $3a + 2d + 2d' + 4d''$ $= 208$

In the year 1866 a very curious observation was made by Mr. Newlands, to the effect that when the elements are arranged in a continuous series in the order of their atomic weights, commencing with hydrogen, there is at equal intervals in the series a recurrence of the same or similar general characters, both physical and chemical. This periodic revival of characteristics occurs, with a few exceptions, at about every eighth member of the series, as will presently be shown. This discovery has been elaborately studied by Mendelejeff and Lothar Meyer, and that which has long been vaguely recognised is now fully established, namely, that the properties of the elements stand in a definite relation to their atomic weights. Further, if we write down a few of the best known elements in the order indicated, and add the formulæ of some of their most characteristic compounds, the reader who has carefully studied their properties will recognise at once the existence of some kind of periodic relation.

Element	Atomic weight	Relative density	Atomic volume[1]	Chief chloride	Chief hydride	Chief oxide
Li	7	·578	12	LiCl	?	Li_2O
Be	9	1·64	5·6	$BeCl_2$	?	BeO
B	11	2·68	4·1	BCl_3	BH_3	B_2O_3
C (Graphite)	12	2·2	5·4	CCl_4	CH_4	CO_2
N	14			?	NH_3	N_2O_5
O	16	Liq. ·99	16	OCl_2	OH_2	
F	19				FH	
Na	23	·97	23	NaCl	?	Na_2O
Mg	24	1·7	14	$MgCl_2$	?	MgO
Al	27	2·6	10	$AlCl_3$	?	Al_2O_3
Si	28	2·49	11	$SiCl_4$	SiH_4	SiO_2
P	31	2·1	14	PCl_3 & PCl_5	PH_3	P_2O_5
S	32	2·07	16	SCl_2 & SCl_4	SH_2	SO_3
Cl	35·5	Liq. 1·38	25		ClH	ClO_2
K	39·1	·865	45	KCl	?	K_2O
Ca	40	1·8	22	$CaCl_2$	?	CaO

[1] The volume of an atom of an uncombined element in a liquid or solid state is assumed to be in direct proportion to the atomic

The gradual waxing and waning of density and atomic volume, of valency, of disposition to combine with chlorine and with hydrogen, and the basylous or acidulous character of the oxides of these elements may be displayed more conspicuously by a graphic construction such as the following.

Here the curve is traced by the intersection of horizontal lines, whose length is proportional to the atomic weights, and vertical lines, which indicate the atomic volumes. It is certainly remarkable that the atomic volume should not increase with the atomic weight in the same or nearly the same ratio, but should thus increase and diminish in value at pretty regular intervals, but the regularity with which these alternations are followed by the various physical and chemical characters of the elements is still more interesting. The student will recognise without difficulty such facts as the following by reference to the diagram.

It will be seen first that, passing from left to right, the ascending parts of the curve are occupied by elements, the electro-negative character of which becomes more and more pronounced the higher they stand, whilst on each summit and the succeeding declivity are stationed elements of electro-positive character which becomes less marked as we go downwards. The elements which are to be found upon the elevated portions of the curve, and which therefore have the largest atomic volume, habitually behave more energetically as chemical agents than those which occupy the lower portions. And this general inference is supported by reference to the heats of formation of the chlorides or oxides of the elements in question. On the ascending portions of the curve, including the topmost points, are also found the volatile and fusible elements, whilst those which occur at the lowest parts are substances which are, for the

weight, and inversely as the density of the substance: hence *atomic volume* is an abbreviated expression meaning the quotient obtained by dividing the atomic weight by the density.

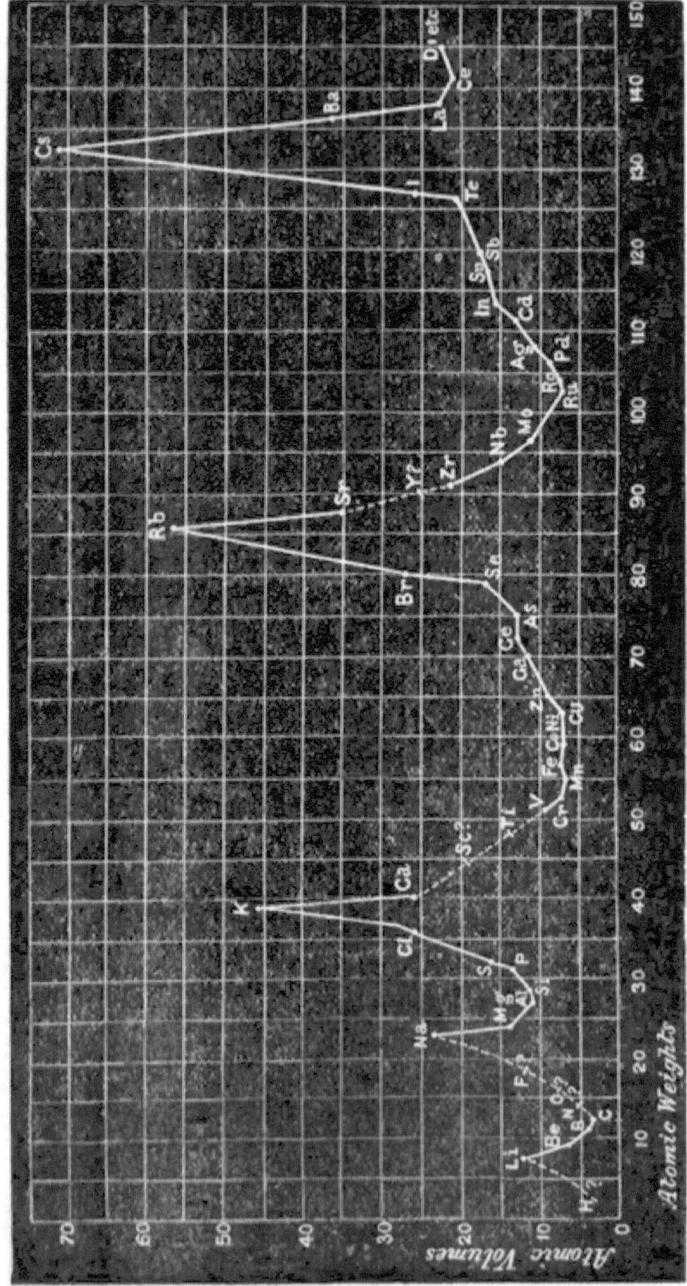

most part, fusible and vaporisable only with difficulty. Malleability and ductility and other physical properties are also exhibited periodically at successive points upon the curve. Moreover, it has been found that the properties of the corresponding compounds of the elements, so far as they can be at present traced, exhibit a similar periodic character. Thus the melting points of the normal chlorides of the elements, when set out graphically in the manner shown in the case of the atomic volumes, give a curve which exhibits the same general characters, consisting of alternate elevations and depressions which correspond roughly with those shown in the diagram. The corresponding bromides and iodides also give similar curves, following pretty closely the curve of the chlorides.

The only character of importance which does not exhibit alternate maxima and minima is specific heat. This, however, though not periodic with increase of atomic weight, is somewhat dependent upon atomic volume, for it will be noticed by reference to the diagram that those elements which do not closely conform to the law of Dulong and Petit have not only a small atomic weight, but small atomic volume.

The table given on p. 292 contains the elements arranged in a manner nearly identical with that proposed by Mendelejeff, and represents a classification based upon the periodic law. It will be noticed that the vertical columns contain closely related elements in two parallel series, the members of which alternate in the value of their atomic weights. The first six horizontal lines are now nearly complete, but there are many gaps lower down. It is interesting to notice that the metals gallium, discovered in 1875, and germanium, in 1886, have been found to fill two places previously vacant, and that they present nearly all the characters belonging to the position which, in virtue of their atomic weights, they hold in the system.

The true places of the rare metals yttrium, lanthanum,

292 *Chemical Philosophy.*

	Monads	Diads	Triads or quasi-triads	Tetrads	Triads or pentads	Diads or Hexads	Monads or Triads	Valency various
1	Li7	G9·1	B011	C12	N14	O16	F19	
2	Na23	Mg24	Al27	Si28	P31	S32	Cl35·5	
3	K39·1	Ca40	Sc44	Ti48	V51·4	Cr52·1	?	{Mn54 / Fe56} {Co59 / Ni58} Cu63·3
4	?	Zn65	Ga69	Ge72	As75	Se79	Br80	
5	Rb85·5	Sr87·6	Y90	Zr89·6	Nb94	Mo95·7	?	{Ru104 / Ro104} Pd106 Ag108
6	?	Cd112	In114	Sn118	Sb120	Te126?	I127	
7	Cs133	Ba137	{Ce141 / La139}	?	?	?	?	
8	?	?	?{Di145 / Tr148}	?	?	?	?	
9	?	?	Er166	Pb207	Ta182	W184	?	{Ir193 / Os199} Pt195 Au197
10	?	Hg200	Tl204	?	Bi208	?	?	
11	?	?	?	Th234	?	U239	?	

cerium, didymium, terbium and erbium are still unsettled, as well as those of ytterbium, decipium, philippium, samarium, &c., the discovery of which has more recently been announced.

This table requires a few remarks. In the first place there are some elements, as, for instance, iron, manganese, cobalt, and nickel, copper, silver, and gold, for which a place cannot readily be found. Manganese is sometimes considered to stand in the same relation to chlorine that chromium does to sulphur. Silver is undoubtedly allied, though not very closely, with sodium, whilst it is also connected with copper on the one hand and with mercury on the other. Gold, again, is unquestionably triad, whilst the platinum metals to which it is most nearly related exhibit even valency. It is also obvious that such relations as are known to exist between thallium and the alkali group, silver and lead, and the special peculiarities of a number of elements, such as mercury, gallium, carbon, and copper, are neither accounted for nor indicated by their position. So that whilst the scheme proposed by Mendelejeff displays the fundamental principle of periodicity very clearly, there are many difficulties in the way of its adoption as a final expression of the true relations subsisting among the elements.

EXERCISES ON SECTION V.

1. What experiments and reasoning would lead you to regard chlorine as a monad element?

2. In what respects does iodine resemble and differ from bromine?

3. Discuss the title of hydrogen to be considered (*a*) a non-metal; (*b*) a metalloid.

4. Classify the solid elements according to their relative densities.

5. What are the properties which distinguish chlorine, and the elements most like chlorine, from potassium, and the elements which are most like potassium?

6. Compare and contrast the elements phosphorus and arsenic, both in their physical and chemical properties.

7. What are the special characteristics of carbon which distinguish it from other elements?

8. Describe as many as you can of the modified forms of carbon, sulphur, and phosphorus.

9. What acids are known consisting of hydrogen, oxygen, and phosphorus? Give their graphic constitutional formulæ, and point out the basicity of each.

10. What differences and analogies exist between the corresponding compounds of potassium and sodium?

11. To what elements does mercury present the greatest resemblance? Point out instances of analogy in properties.

12. Show by a comparison of their compounds in what respects the elements carbon, silicon, tin, and lead may be said to constitute a natural group.

13. Give the formulæ, simple and constitutional, of the following compounds: common salt, caustic potash, sulphuric acid, baryta, barium peroxide, lead peroxide, iron peroxide, silver phosphate, silver arsenate, silver arsenite, kakodyl, and tartar-emetic.

14. Give a full account of the two oxides of nitrogen known as nitric oxide and nitric peroxide respectively.

15. Give instances of compounds which do not comply with the law of even numbers.

16. What are the chief points of resemblance and difference between the corresponding calcium and magnesium compounds?

17. Assign a place among the elements to each of the following bodies: copper, nitrogen, boron, thallium, vanadium, gold. Give full reasons in each case.

18. The constitution of phosphorus compounds has been explained on the assumption of the trivalence of the phosphorus atom. Discuss this view.

19. By what arguments would you support the number 32 against the number 16 as the atomic weight of sulphur?

20. Caustic potash is often said to be formed upon the water type. Explain this.

21. What is an alum? At the temperature of 180° common alum loses $\frac{23}{24}$ of its water. How may this fact be brought to bear upon the construction of its formula?

22. Classify some of the better known elements according to their valency, mentioning and explaining, as far as possible, doubtful cases.

23. Write the formulæ of orthosulphuric acid and some of its salts. An ortho-acid may be defined as one in which the characteristic element (sulphur in this case) is saturated with hydroxyl.

24. One gram of phosphorus yields 2·2903 grams of phosphoric anhydride. Will this enable you to determine the atomic weight of the element? and, if not, what further data are necessary?

25. How much phosphorus is contained in 120 lbs. of bone-ash, consisting of $Ca_3(PO_4)_2$ 88·5 parts and $CaCO_3$ 11·5 parts in 100?

26. What weight of phosphorus is contained in 10 litres of phosphorus vapour at 1040°? Pressure normal.

27. What weight of phosphorus is contained in 10 litres of phosphine at 12°? Pressure normal.

28. Write the formulæ of a few 'molecular' compounds and consider in each case whether the compound can be reasonably represented by a single unitary formula.

29. Calculate the percentage composition of cryolite, Na_3AlF_6.

30. How many kilograms of litharge can be obtained from 0·5 kilos. of lead, and what volume of oxygen measured at 20° and under 765 mm. is absorbed in the process?

SECTION VI.

CHAPTER XXVII.

CLASSIFICATION OF COMPOUNDS.

ACIDS, BASES, SALTS.

Acids.—The only element common to all acids is hydrogen, and the characteristic property of this hydrogen is its ready exchangeability for a metal. It does not appear that this is due to any peculiarity of the hydrogen itself, but to the fact that, in all well-defined acids, it is associated with other elements of very eager affinities. Acids may be divided into two classes according as oxygen does or does not enter into their composition. A classification of acids according to the nature of their constituent elements was given on p. 229.

It is important to remark, however, that, although such a distinction may for certain purposes be occasionally recognised, there is no essential difference between acids which contain oxygen and those which do not.

Now, it is obvious that the activity with which hydrochloric acid, for example, produces salts when in contact with suitable metals, is due chiefly to the elective attraction of the chlorine for the metal, and that hydrocyanic acid, a molecule of which contains the same quantity of hydrogen, does not act upon metals with the same promptitude, because the cyanogen group has not the same attraction for metals as chlorine.

The chemical energy of an acid, then, does not depend

upon the hydrogen which it contains, but upon the electronegative character, more or less strongly marked, in the elements with which the hydrogen is associated. It is a very remarkable fact, however, that the existence of hydrogen in a body, in the position which it occupies in acids, namely, in the position of a base, does confer upon the compound one very general characteristic, and that is a sour taste. The sourness of soluble hydrogen salts would, perhaps, have been considered no more remarkable than the astringency of iron salts, the bitterness of magnesium, or the sweetness of beryllium compounds, but that it constitutes a character which serves to differentiate real acids from bodies such as ammonia, alcohols, and metallic hydrates, which all resemble acids so far as to exchange, under certain circumstances, part of their hydrogen for metals.

Oxyacids are bodies of very varied character. Some considerations concerning them have already been dwelt upon in the article on Oxygen (Chap. XXIII., p. 229).

With regard to the properties of acids in general, sufficient has already been written in the article on Hydrogen (Chap. XXIV., p. 245). It now only remains, therefore, to discuss that property of acids known as 'basicity.' This inquiry amounts in some cases to a determination of the weight of the molecule, and reference has already been made to an example of the kind in the article on Molecular Weights (Chap. XXV., p. 131). The question is, however, one of great importance, and deserves to be examined a little more in detail. The basicity of an acid depends upon the number of atoms of hydrogen directly exchangeable out of each molecule for metallic atoms ; an acid containing one such atom of hydrogen being called monobasic, one containing two such atoms is dibasic, or three tribasic, and so on. We may infer the basicity of a given acid from the number of distinct salts it is capable of yielding with each metal. Monobasic acids generally give one salt, dibasic acids, as shown in the case of sulphuric acid, give two, and tribasic acids give three. Mere inspection of the formula of an acid does not, however, furnish sufficient information upon which to form a judgment regarding the basicity of the acid and the number of salts it is capable of producing; for it by no means follows, from the existence of two or three atoms of hydrogen in the molecule, that it should be dibasic or tribasic. The basicity depends not upon the total number of hydrogen atoms present, but upon the total number endowed with this particular power of metallic exchange.

The three following acids of phosphorus afford an example of this. Each molecule contains three atoms of hydrogen, but the first is monobasic, the second dibasic, the third tribasic :—

    Hypophosphorous acid   . H_3PO_2 or HPH_2O_2
    Phosphorous acid   .   . H_3PO_3 or H_2PHO_3
    Phosphoric acid   .   . H_3PO_4

In this and similar cases it has been generally observed that increase of basicity, as well as of sourness and general chemical acidity, accompanies the addition of oxygen. Hence it has been inferred that those atoms of hydrogen which are exchangeable for metals are combined more intimately with oxygen than others which are not similarly exchangeable. On this hypothesis the formulæ of the three acids referred to above are represented in the following manner :—

$$\text{Hypophosphorous acid} \quad . \quad O=P{\Large\diagdown}\!\!\!\!{\Large\diagup}\!\!\!\begin{matrix}H\\H\end{matrix}\,-OH$$

$$\text{Phosphorous acid} \quad . \quad O=P\begin{matrix}H\\ \\OH\end{matrix}\diagup\!\!\!\!\diagdown OH$$

$$\text{Phosphoric acid} \quad . \quad O=P\diagup\!\!\!\!\diagdown\begin{matrix}OH\\-OH\\OH\end{matrix}$$

But acids are capable of undergoing a variety of modifications under the influence of reagents, and some of these are available as evidence towards the establishment of their basicity. It has been found, for example, that acids which possess only one basic atom of hydrogen, or are monobasic, yield only one amide and one chloride. Acetic acid, for example, gives the following derivatives, the relation of which to acetic acid is obvious from the formulæ—

Acetic acid	$HO.C_2H_3O$
Acetyl chloride	$Cl.C_2H_3O$
Acetamide	$NH_2.C_2H_3O$.

Dibasic acids give two derivatives of the same kind, and tribasic acids give three.

We may now resume the case of sulphuric acid, adverted to in Chap. XV., as a very instructive example.

The simplest formula for sulphuric acid is H_2SO_4. The problem before us is to prove that this formula, with the relative weight 98, is the formula of the molecule, and that the acid is dibasic.

1. The fact of the existence of two classes of sulphates has been already appealed to. It is clear that the whole of the hydrogen of the acid is replaceable by metals, and that it is capable of replacement one half at a time. We arrive then at such formulæ as these :—

$$\left.\begin{matrix}Na\\H\end{matrix}\right\}SO_4 \qquad \left.\begin{matrix}Na\\Na\end{matrix}\right\}SO_4 \qquad \left.\begin{matrix}K\\K\end{matrix}\right\}SO_4 \qquad \left.\begin{matrix}K\\Li\end{matrix}\right\}SO_4$$

Evidence of this kind is almost conclusive, but it happens that in the present instance other testimony is abundant.

2. Thus we might show that sulphuric acid is produced when sulphur trioxide and water, both bodies of known molecular weight, unite together—

$$SO_3 + OH_2 = SO_4H_2.$$

3. Also when sulphur dioxide is dissolved in water and the solution exposed to contact with oxygen—

$$SO_2 + OH_2 + O = SO_4H_2.$$

4. It is also formed by the direct union of hydrogen peroxide and sulphur dioxide—
$$SO_2 + O_2H_2 = SO_4H_2.$$

A sulphate being generated in the following strictly parallel case of combination—
$$SO_2 + PbO_2 = SO_4Pb.$$

5. Sulphuric acid is generated by the decomposition of sulphuryl chloride by water—
$$SO_2Cl_2 + 2H_2O = SO_4H_2 + 2HCl.$$

According to equations 2, 3, 4, the whole of the acting materials enter into the composition of the sulphuric acid which is produced, and therefore there is a strong probability in favour of the molecular weight of sulphuric acid being the sum of their separate molecular weights. Similar considerations apply to equation 5. The molecule of sulphuric acid is here shown to be built up of the molecules of the sulphur oxychloride and water, minus two atoms of hydrogen and two atoms of chlorine.

6. We now come to some reactions in which various constituents of the sulphuric acid molecule are replaced. When the acid is heated with phosphoric chloride, two successive reactions of the same kind ensue. In the first an atom of chlorine is introduced in place of the group (OH) which is removed.

$$H_2SO_4 + PCl_5 = HClSO_3 + POCl_3 + HCl.$$

And this kind of exchange is repeated in the second reaction:

$$HClSO_3 + PCl_5 = Cl_2SO_2 + POCl_3 + HCl.$$

The relation of the two new chlorides to sulphuric acid may be exhibited most satisfactorily by writing the formulæ in the following manner :—

$$\begin{matrix} HO \\ HO \end{matrix} SO_2 \qquad \begin{matrix} HO \\ Cl \end{matrix} SO_2 \qquad \begin{matrix} Cl \\ Cl \end{matrix} SO_2$$

Sulphuric acid or Hydrate.   Sulphuryl chlor-hydrate.   Sulphuryl chloride.

7. Two 'amides,' standing in the same kind of relationship towards sulphuric acid, also exist. Their formulæ may be written thus :—

$$\begin{array}{ccc} \mathrm{HO}\!\!>\!\!\mathrm{SO}_2 & \mathrm{HO}\!\!>\!\!\mathrm{SO}_2 & \mathrm{NH}_2\!\!>\!\!\mathrm{SO}_2 \\ \mathrm{HO} & \mathrm{NH}_2 & \mathrm{NH}_2 \\ \text{Sulphuric acid.} & \text{Sulphamic acid.} & \text{Sulphamide.} \end{array}$$

The existence of these compounds and the corresponding chlorides supplies evidence similar to that deducible from a knowledge of the acid and double sulphates (1). These facts tend to show that the hydrogen contained in a molecule of sulphuric acid is divisible into two equal parts, and that these two hydrogen atoms are probably more closely united with two atoms of oxygen than with other constituents of the molecule.

8. When concentrated sulphuric acid is made to act on certain hydrocarbons, alcohols, phenols, and other bodies, compounds are produced which have the properties of monobasic acids. These compounds are in some degree analogous to salts; at any rate, sulphuric acid could not produce derivatives of this kind if it did not contain at least two atoms of basylous hydrogen.

$$\underset{\text{Benzene.}}{C_6H_6} + \underset{\text{Sulphuric acid.}}{H_2SO_4} = \underset{\text{Water.}}{H_2O} + \underset{\text{Benzene-sulphonic acid.}}{C_6H_5(SO_3H)}$$

$$\underset{\text{Ethylic alcohol.}}{C_2H_5OH} + \underset{\text{Sulphuric acid.}}{H_2SO_4} = \underset{\text{Water.}}{H_2O} + \underset{\text{Ethyl-sulphuric acid.}}{C_2H_5O(SO_3H)}$$

$$\underset{\text{Phenol.}}{C_6H_5OH} + \underset{\text{Sulphuric acid.}}{H_2SO_4} = \underset{\text{Water.}}{H_2O} + \underset{\text{Phenol-sulphonic acid.}}{C_6H_4(OH)(SO_3H)}$$

All the foregoing facts point to the same conclusion, and it matters not whether we write the formula of sulphuric acid—

$$H_2SO_4 \text{ or } \genfrac{}{}{0pt}{}{H}{H}\!\!\bigg\}SO_4 \text{ or } (HO)_2SO_2 \text{ or } H_2O.SO_3,$$

in the attempt to record one or other of its various modes of formation or decomposition. These different expressions represent one and the same molecular weight.

In one respect, which has not yet been referred to, there is a difference between acids which contain oxygen and those which do not contain that element. Compounds in which there is no oxygen cannot by any possibility be made to yield water. Hence there are no anhydrides corresponding to the haloid acids. Oxyacids, however, may, in various ways, be made to furnish water and an oxide, which is called an *anhydride*. Thus sulphurous acid splits up readily in the following manner :—

$$\underset{\text{Sulphurous acid.}}{H_2SO_3} = \underset{\text{Water.}}{H_2O} + \underset{\text{Sulphurous anhydride.}}{SO_2}$$

Dibasic acids give their anhydrides most readily when simply heated, whereas monobasic acids generally require to be treated with some dehydrating agent. This may, perhaps, be connected in some way with the fact that a dibasic acid always contains within itself the elements of water, whilst a monobasic acid cannot generate water except by the combined action of two molecules.

Dibasic.

$$\underset{\text{Carbonic acid.}}{H_2CO_3} = H_2O + \underset{\text{Anhydride.}}{CO_2}$$

$$\underset{\text{Succinic acid.}}{H_2C_4H_4O_4} = H_2O + \underset{\text{Anhydride.}}{C_4H_4O_3}$$

Monobasic.

$$\underset{\text{Nitric acid.}}{\left.\begin{matrix}HNO_3\\HNO_3\end{matrix}\right\}} = H_2O + \underset{\text{Anhydride.}}{N_2O_5}$$

In the case of some monobasic acids, the anhydride can only be obtained by a succession of operations. In order to prepare acetic anhydride, for example, the chloride is first made from the acid, and this compound is then allowed to react upon a salt.

$$\underset{\text{Acetic chloride.}}{C_2H_3OCl} + \underset{\text{Sodium acetate.}}{C_2H_3O.ONa} = \underset{\substack{\text{Sodium}\\\text{chloride.}}}{NaCl} + \underset{\substack{\text{Acetic oxide or}\\\text{anhydride.}}}{(C_2H_3O)_2O}$$

Definition of 'Base.'

This difference between monobasic and dibasic acids is, however, not sufficiently general to be maintained as a distinction of much practical importance.

Bases.—The idea implied by the word 'base' belongs to the obsolete dualistic theory of salts. According to this theory, every salt was supposed to be made up of two parts, one of which, consisting of a metallic oxide, formed the base; whilst the other part, usually the oxide of a non-metal, or some group consisting of carbon, hydrogen, and oxygen, was regarded as the acid. Sulphate of soda, for example, was, according to dualistic phraseology, composed of the base soda or oxide of sodium, and sulphuric acid or trioxide of sulphur, and its formula was written $NaOSO_3$, or, adopting modern atomic weights, Na_2OSO_3.

Now, although this idea has been considerably modified, and we no longer look upon a salt as a double structure of this kind, the fact remains that those oxides of metals which are not overcharged with oxygen do agree in the property of uniting or reacting with acids, so as more or less completely to neutralise them. And this property is shared by the hydrates corresponding with these oxides. Adopting the old name, and taking this neutralising faculty as the criterion of the basic function of oxides, we may define a base as *a metallic oxide or hydroxide capable of saturating acids.*

It will be noticed that a basic hydroxide and basic oxide have the same relation to each other as an acid and its anhydride, and consequently it would be quite reasonable to speak of the anhydrous oxide as an anhydride, reserving the term base for the hydroxide. For example—

$2HNO_3$ — OH_2 = N_2O_5
Acid.   Water.   Acid anhydride.

$2KHO$ — OH_2 = K_2O
Basic hydroxide or base.   Basic oxide or basic anhydride.

$$\underset{\text{Acid}}{H_2SO_4} - \underset{\text{Water.}}{OH_2} = \underset{\text{Acid anhydride.}}{SO_3}$$

$$\underset{\text{Basic hydrate or base.}}{Ca(HO)_2} - \underset{\text{Water.}}{OH_2} = \underset{\text{Basic oxide or basic anhydride.}}{CaO}$$

It must not be forgotten that ammonia and a certain class of its derivatives also possess the power, in a remarkable degree, of uniting with acids and forming perfectly neutral compounds, which present all the external characteristics, and many of the chemical properties, of ordinary salts. Ammonia and the alkaloids are therefore commonly regarded as bases, though it is well to remind the student that these bodies differ from metallic oxides or hydroxides in their reactions, both with acids and with anhydrides. Thus with acids, ammonia enters into direct combination, without the formation of any secondary product, thus:—

$$NH_3 + HNO_3 = NH_3HNO_3$$
$$\text{or } NH_4NO_3$$

Whereas basic oxides or hydroxides invariably generate water in addition to the salt.[1]

$$\underset{\text{Base.}}{KHO} + \underset{\text{Acid.}}{HNO_3} = \underset{\text{Salt.}}{KNO_3} + \underset{\text{Water.}}{H_2O}$$

Again, ammonia combines with acid anhydrides generally, but the compounds which result are not salts, whilst a basic oxide under the same circumstances always yields a salt of the ordinary type. For example:—

$$BaO + SO_3 = BaSO_4$$

Such a reaction is obviously the counterpart of that which occurs when an acid or hydrogen salt is formed from water and the anhydride.

$$H_2O + SO_3 = H_2SO_4$$

[1] The demonstration of this fact is very generally neglected by teachers. An easy experiment by which an appreciable quantity of water may be collected in a few minutes consists in passing dry hydrochloric acid over gently-heated litharge.

Salts.

The general property of neutralising acids is associated, in the case of all those bases which are soluble in water, with a peculiar soapy taste, and with a power of affecting vegetable colours. The reds are changed to blue or green, the blues to green, and the yellows to brown or brownish red. In litmus the alteration of tint characteristic of alkaline reaction is the result of the conversion of the feeble litmus acid (red) into a blue salt. Somewhat similar changes probably occur in other cases.

Salts.—It has been stated that an acid is a body in which hydrogen is united with a strongly electro-negative element or group of elements. When such a body is acted upon by a metal or metallic oxide, the hydrogen is exchanged for a metal, or something equivalent to it, and a salt is formed. We may define a salt, then, as a compound in which a metallic or other radicle of like nature is combined with a radicle of more or less decided electro-negative character. It has already been explained that electro-negative elements are those which in the electrolysis of the salt are collected at the positive electrode, and which by uniting with hydrogen form acids. In short, the radicles which, in union with hydrogen, form acids, and with metals form salts, are always composed either of the non-metallic, or, as they are called in this book, oxygenic elements, or of certain metals in company with a relatively large quantity of oxygen.

A considerable number of salts are known in which a part of the acid radicle of the normal compound is replaced by oxygen or hydroxyl. Such compounds are often called basic salts. The following are some examples which serve to show the general character of these compounds :—

$$\text{Plumbic hydroxychloride} \quad \text{Pb} \begin{cases} \text{OH} \\ \text{Cl} \end{cases}$$

$$\text{Plumbic hydroxynitrate} \quad \text{Pb} \begin{cases} \text{NO}_3 \\ \text{OH} \end{cases}$$

Bismuthic hydroxynitrate $\quad Bi \begin{cases} NO_3 \\ NO_3 \\ OH \end{cases}$

Ferric oxysulphate $\quad Fe_2 \begin{cases} (SO_4)_2 \\ O \end{cases}$

Salts may be formed in a variety of ways, amongst which are the following :—

(*a*) Union of elements,
$$Na_2 + Cl_2 = 2NaCl.$$

(*b*) Union of acid and basic oxide,
$$CaO + CO_2 = CaCO_3.$$

(*c*) Action of metal on acid,
$$Zn + H_2SO_4 = ZnSO_4 + H_2.$$

(*d*) Action of metallic oxide or hydroxide on acid,
$$ZnO + H_2SO_4 = ZnSO_4 + OH$$
$$NaHO + HNO_3 = NaNO_3 + OH_2$$

(*e*) Double decomposition between two salts,
$$BaCl_2 + Na_2HPO_4 = 2NaCl + BaHPO_4.$$

As regards the properties of salts no very general statements can be made. Liquid and solid, soluble and insoluble, crystalline and uncrystallisable bodies are alike included under this denomination, so that no comparison can now be instituted between the bodies thus named by the chemist and that substance which was probably the prototype of all salts, namely, *common salt* (Gr. ἅλς, Lat. *sal*, Fr. *sel*, Ger. *salz*).

In the production of salts the potential energy of the reacting materials is more or less exhausted, so that salts are generally bodies of feeble attractions. We observe this in the comparatively small number of compounds into which entire salts enter, and the facility with which such combinations are broken up.

CHAPTER XXVIII.

DERIVATIVES OF AMMONIA.

Ammonium Theory.—Ammonia is excessively soluble in water, and the solution has all the properties of a powerful alkali. Like caustic potash, it restores the blue colour to litmus which has been reddened by an acid, it absorbs carbonic acid from the air, it saturates acids, forming a class of crystallisable neutral salts, which are isomorphous with the corresponding salts of potassium. Like potash, also, it precipitates hydroxides when added to the solutions of a great many metallic salts, though in a few cases, *e.g.* with mercuric salts, it gives rise to compounds of a different character. Lastly, solution of ammonia differs from solution of caustic potash in that when evaporated it leaves no residue. In consequence of this want of stability, solution of ammonia is often referred to as the *volatile alkali*.

Guided by this marked resemblance between the compounds of ammonia and those of potassium, chemists have been led to regard both series as constituted in the same manner. Writing down in parallel columns a list of ammoniacal and potash salts, and eliminating from each pair of formulæ the symbols which are common to them both, we find that the group composed of one atom of nitrogen and four atoms of hydrogen, NH_4, is the residue left over in the one case, whilst the solitary symbol K remains in the other. The latter represents the radicle potassium, which, when isolated, presents all the characteristics of a metal. The conclusion seems almost inevitable that the radicle NH_4 would also exhibit metallic lustre, conductivity, and so forth, if it could be obtained in a free state. In anticipation of this hypothesis being confirmed, it has been called *ammonium*.

Compounds of Ammonium.			Potassium.
$NH_3.H_2O$	or NH_4HO	or $AmHO$	KHO
$NH_3.HCl$	or NH_4Cl	or $AmCl$	KCl
$NH_3.HNO_3$	or NH_4NO_3	or $AmNO_3$	KNO_3
$NH_3.H_2SO_4$	or NH_4HSO_4	or $AmHSO_4$	$KHSO_4$
$(NH_3)_2.H_2SO_4$	or $(NH_4)_2SO_4$	or Am_2SO_4	K_2SO_4
$NH_3H_3PO_4$	or $(NH_4)H_2PO_4$	or AmH_2PO_4	KH_2PO_4
$(NH_3)_2H_3PO_4$	or $(NH_4)_2HPO_4$	or Am_2HPO_4	K_2HPO_4
$(NH_3)_3H_3PO_4$	or $(NH_4)_3PO_4$	or Am_3PO_4	K_3PO_4

A curious phenomenon occurs when a lump of sodium amalgam is placed in a solution of sal ammoniac, or when the same salt, dissolved in water, is submitted to electrolysis, a globule of mercury being employed as the negative electrode. The mercury swells enormously, becoming so light as to float in the solution, without at the same time losing altogether its metallic aspect. In a few minutes, however, the spongy mass again contracts, hydrogen and ammonia make their escape, and the mercury recovers its original volume and lustre. The body thus produced was long regarded as a true compound of mercury with the hypothetical metallic radicle, and as such received the name *ammonium amalgam*. It has been shown, however, that the mercury whilst in this condition yields to pressure in the same way as froth, and it is doubtful whether the phenomenon is really due to the formation of an amalgam, or to the mechanical distension of the mercury by gas.

The radicle ammonium has, in fact, never been isolated, and there is no great probability of its ever being obtained. The ammonium theory finds a sufficient basis in the isomorphism of the ammonia and potash salts, and the general resemblance of the hydrate and its derivatives (to be mentioned presently) to the fixed alkalis, and the theory itself should be regarded merely as a summary and memorandum of these relations.

Amines.—The hydrogen of ammonia admits of substitu-

tion in three successive stages, and the compounds thus formed are believed to be constituted in the same manner as ammonia itself, or, using the ordinary expression, they belong to the ammonia type,
$$N \begin{cases} H \\ H \\ H \end{cases}$$
the replacing radicles being severally combined with the nitrogen in the same manner as the hydrogen atoms of the type.

When positive radicles, consisting of metals or hydrocarbons, are substituted for the hydrogen in the ammonia molecule, the resulting compounds retain the alkalinity and general basic character of ammonia. Such compounds are called *amines*. They generally combine with acids, but are unaffected by treatment with alkalies. There are several classes of amines resulting from the more or less extensive replacement of the hydrogen or the coalescence of two or more molecules of ammonia. Their constitution will be understood after inspection of the following table, in which the symbol R represents a positive radicle, the valency of which is indicated by dashes :—

Monamines.	Diamines.	Triamines.
Primary. $N \begin{cases} R' \\ H \\ H \end{cases}$	Primary. $N_2 \begin{cases} R'' \\ H_2 \\ H_2 \end{cases}$	Primary. $N_2 \begin{cases} R''' \\ H_3 \\ H_3 \end{cases}$
Secondary. $N \begin{cases} R' \\ R' \\ H \end{cases}$	Secondary. $N_2 \begin{cases} R'' \\ R'' \\ H_2 \end{cases}$ $N_2 \begin{cases} R'' \\ R'_2 \\ H_2 \end{cases}$	Secondary. $N_3 \begin{cases} R''' \\ R''' \\ H_3 \end{cases}$ $N_3 \begin{cases} R''' \\ R'_3 \\ H_3 \end{cases}$, &c.
Tertiary. $N \begin{cases} R' \\ R' \\ R' \end{cases}$	Tertiary. $N_2 \begin{cases} R'' \\ R''_2 \\ R'' \end{cases}$ $N_2 \begin{cases} R'' \\ R''_2 \\ R'_2 \end{cases}$ $N_2 \begin{cases} R'' \\ R'_2 \\ R'_2 \end{cases}$	Tertiary. $N_3 \begin{cases} R''' \\ R''_3 \\ R''' \end{cases}$ $N_3 \begin{cases} R''' \\ R''N_3 \\ R'_3 \end{cases}$ $N_3 \begin{cases} R''' \\ R'_3 \\ R'_3 \end{cases}$ &c.

Monamines are capable of saturating only one molecule of a monobasic acid, but diamines unite with two, and triamines with three molecules of such acids. The diamines and triamines also yield intermediate compounds. Representing the monamines, diamines, and triamines respectively by the symbols Am, Am^2, and Am^3, the general formulæ of the chlorides may be written in the following manner:

$$Am.HCl \qquad Am^2.HCl \qquad Am^3.HCl.$$
$$ Am^2.2HCl \qquad Am^3.2HCl.$$
$$ Am^3.3HCl.$$

The tertiary amines unite with iodides and other salts of alcohol radicles in the same manner as with acids, furnishing compounds which may be formulated either as ammonia compounds or as derived from the hypothetical ammonium radicle. Thus triethyl-amine combines with ethyl iodide, forming a crystalline compound, which may be represented either as $N(C_2H_5)_3.C_2H_5I$, or, more consistently and conveniently, as tetrethyl-ammonium iodide $N(C_2H_5)_4I$. This compound is obviously the analogue of ammonium iodide. When acted upon by silver oxide and water, it exchanges iodine for hydroxyl, and gives rise to a hydroxide of tetrethyl-ammonium, $N(C_2H_5)_4HO$, a compound which may be obtained in the solid state, and which in its causticity and alkalinity closely resembles potassic hydroxide. The existence of this compound, and others of similar nature, furnishes strong evidence in favour of the ammonium theory. Like their hypothetical prototype, however, the substituted ammonium radicles $N(C_2H_5)_4$, &c., are not known in the free state.

A question has at different times been raised regarding the constitution of ammonium compounds. It seems, however, to be generally admitted that they contain pentad nitrogen, ammonium chloride being represented graphically by the formula—

Constitution of Ammonium Salts.   311

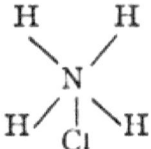

and other ammonium compounds in a corresponding manner. Experiment has been brought to bear on the subject in the following manner :

Methyl-triethyl ammonium chloride may be represented either as an ammonium compound—

$$\begin{array}{c} C_2H_5 \quad C_2H_5 \\ \diagdown \quad \diagup \\ N \\ \diagup \ | \ \diagdown \\ C_2H_5 \ Cl \ CH_3 \end{array}$$

or, as formed on the type of sal-ammoniac, regarded as ammonia hydrochloride, $NH_3.HCl$. If the latter view is correct, there ought to be two isomeric compounds having the formulæ—

$$N(C_2H_5)_3.CH_3Cl \text{ and } N(C_2H_5)_2(CH_3).C_2H_5Cl.$$

It is found, however, that the result of bringing together triethylamine and methyl chloride, is a body identical in every respect with the product obtained by combining methyl-diethylamine with ethyl chloride.

According to the ammonium formula, which these results tend to support, each radicle is in the same position with respect to the nitrogen, and no isomerism is possible.

Concerning the formation of sal-ammoniac, however, it has been justly remarked, that if we adopt the formula $N^v(H_4Cl')$, it is difficult to explain why, in the combination of hydrochloric acid and ammonia, the chlorine should be represented as leaving the hydrogen, for which it has great attraction, in order to link itself with nitrogen, for which it appears to have little attraction.

$$\begin{array}{c} H \\ | \\ Cl \end{array} + \begin{array}{c} H \\ \diagup \\ N-H \\ \diagdown \\ H \end{array} = \begin{array}{c} H \quad H \\ \diagdown \diagup \\ N-H \\ \diagup \diagdown \\ Cl \quad H \end{array}$$

The truth is probably this, that it is the hydrogen of the ammonia, and not the nitrogen, which attracts the chlorine. In other words, it is the entire molecule NH_3 which attracts and combines with the entire molecule HCl. As the result of such combination a new molecule is formed in which the atoms of nitrogen, hydrogen, and chlorine are bound together in a closer and more intimate union than it is possible to express by symbols which assume a limited combining capacity for each element. But this is only one of many cases which show how incompetent are all existing systems of formulæ to represent adequately the essential nature of bodies and the chemical changes they undergo.

Phosphines, Arsines, Stibines.—By replacing the hydrogen of phosphoretted, arsenetted, and antimonetted hydrogen by hydrocarbon groups, extensive series of bases are obtained, which have received the above names, and, save for their extreme oxidability, agree in general characters with the compound ammonias.

Amides.—When the hydrogen of ammonia is replaced by oxidised or other negative radicles, compounds called *amides* are formed. These bodies are, generally speaking, neutral, though some which contain the radicles of weak acids are still capable of entering into combination with acids, and a few are even very decidedly basic; urea or carbamide, $CO(NH_2)_2$ for example.

On the other hand, some of the amides derived from

polybasic acids are themselves acid bodies. This has been already referred to (Acids, p. 301).

The general formulæ of amides correspond with those of amines, and they may be classified into primary, secondary, and tertiary monamides, diamides, and triamides, and, substituting negative R or $\overline{\text{R}}$ for R in the formulæ given on p. 309, they may be tabulated in the same way as the amines. One slight addition is necessary. Monamines containing bivalent radicles are not known, but monamides in which two atoms of hydrogen are replaced by one bivalent acid radicle, are known under the name imides.

Cyanic acid or Carbimide.   Succinimide.

$$N \begin{cases} (CO)'' \\ H \end{cases} \qquad N \begin{cases} (C_4H_4O_2)'' \\ H \end{cases}$$

Tertiary derivatives of ammonia, called nitriles or cyanides, are known in which the three atoms of hydrogen are replaced by trivalent hydrocarbon groups. We have, for example,

Formonitrile   Acetonitrile   Benzonitrile
$N(CH)'''$ $N(C_2H_3)'''$ $N(C_7H_5)'''$
or Hydrogen cyanide.   or Methyl cyanide.   or Phenyl cyanide.

These bodies are not basic, and are decomposed, like all amides, by boiling with alkalis. Ammonia is evolved, and a salt formed corresponding with the acid from the ammonium salt of which they are derived by dehydration. Notwithstanding, therefore, that the replacing radicle contains no oxygen, these bodies should be ranked with the amides. The isocyanides (pp. 196 and 238) are basic, and are therefore called carb*amines*.

The residual hydrogen of primary and secondary amides may be replaced by positive or hydrocarbon radicles. Compounds intermediate between amines and amides then result. The following are examples :—

$$N\begin{cases}C_2H_5\\H\\H\end{cases}\quad N\begin{cases}C_2H_5\\C_2H_5\\H\end{cases}\quad N\begin{cases}C_2H_5\\C_2H_5\\C_2H_5\end{cases}$$

Ethylamine.   Diethylamine.   Triethylamine.

$$N\begin{cases}C_2H_3O\\H\\H\end{cases}\quad N\begin{cases}C_2H_3O\\C_2H_3O\\H\end{cases}\quad N\begin{cases}C_2H_3O\\C_2H_3O\\C_2H_3O\end{cases}$$

Acetamide.   Diacetamide.   Triacetamide.

$$N\begin{cases}C_2H_3O\\C_2H_5\\C_2H_5\end{cases}\quad N\begin{cases}C_2H_3O\\C_2H_3O\\C_2H_5\end{cases}\quad N\begin{cases}C_2H_3O\\C_2H_5\\H\end{cases}$$

Diethylacetamide.   Ethyldiacetamide.   Ethylacetamide.

CHAPTER XXIX.

CARBON COMPOUNDS.

BODIES having the same chemical functions, and constituting a series in which, between any two contiguous terms, there is a constant difference of CH_2, are said to be homologous, and the series is a *homologous series*. Relations of a like nature have not hitherto been observed among the compounds of any element but carbon.[1] This peculiarity seems to be connected with a special faculty with which carbon atoms appear to be endowed, the power, namely, of uniting with one another without the intervention of any other element. Thus a very large number of different compounds containing carbon and hydrogen, either alone or associated with oxygen or nitrogen, have already been produced, and a still larger number of possible combinations of carbon is indicated by theory.

By reason of their composition being unknown or their reactions imperfectly studied, many carbon compounds

---

[1] And probably silicon. Unless, indeed, we choose to regard such a pair of compounds as tetrachloride and sesquichloride of manganese, $MnCl_4$, Mn_2Cl_6, as the first two terms of a series Mn_nCl_{2n+2}, the higher members of which are unknown.

remain unclassified, but the number of these has been greatly reduced by modern researches. The following are the most important classes of carbon compounds :—

I. HYDROCARBONS AND HALOID DERIVATIVES.

Some hundreds of compounds consisting only of carbon and hydrogen have been described, but no such compound is known of which two volumes of the vapour contain a larger quantity of hydrogen than is represented by the general formula C_nH_{2n+2}. This fact is explained by the following hypothesis. A single atom of carbon has four units of combining capacity, each of which is capable of being saturated by one atom of hydrogen. We thus arrive at the formula

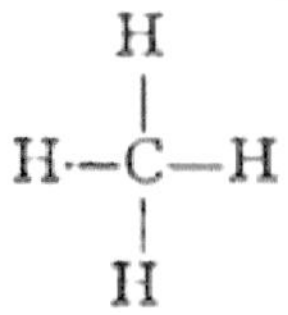

$$H-\underset{\underset{H}{|}}{\overset{\overset{H}{|}}{C}}-H$$

which represents marsh-gas. This body is absolutely saturated, and nothing will induce it to take up additional elements of any kind, except on condition of removing one or more of its hydrogen atoms. If, now, we imagine another atom of carbon similarly loaded with hydrogen presented to it, the only mode in which they can unite together is by the severance of an atom of hydrogen from each group, as represented in the following diagram :—

$$H-\underset{\underset{H}{|}}{\overset{\overset{H}{|}}{C}}-H \quad\quad H-\underset{\underset{H}{|}}{\overset{\overset{H}{|}}{C}}-H$$

or

$$H-\underset{\underset{H}{|}}{\overset{\overset{H}{|}}{C}}-\underset{\underset{H}{|}}{\overset{\overset{H}{|}}{C}}-H$$

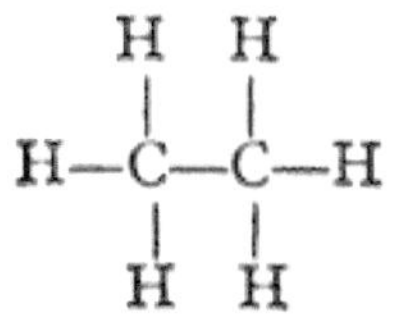

The same process may be repeated *ad infinitum*. The number of hydrogen atoms in the first term of the series being 4, in the second $2 \times 4 - 2$, in the third $3 \times 4 - 4$, in the nth term the number will be $4n - (2n-2)$, or $2n+2$.

The most important series of hydrocarbons at present known are represented by the following general formulæ :—

Paraffins, or Marsh Gas Series,	C_nH_{2n+2}
Olefines, or Olefiant Gas Series,	C_nH_{2n}
Acetylene Series,	C_nH_{2n-2}
Terpene Series,	C_nH_{2n-4}
Benzene Series,	C_nH_{2n-6}
Naphthalene Series,	C_nH_{2n-12}
Anthracene Series,	C_nH_{2n-18}.

From what has already been said as to the manner in which the carbon atoms in a carbon compound become linked together, it is obvious that these general formulæ admit of development in a variety of ways, so as to give rise in each series to a considerable number of isomerides. Thus, if we take the formula for the paraffins and assign different values to n we arrive at such results as the following :—

CH_4	$n = 1$
$CH_3 - CH_3$	$n = 2$
$CH_3 - CH_2 - CH_3$	$n = 3$
$CH_3 - CH_2 - CH_2 - CH_3$	$n = 4$
&c.	&c.

Now it is obvious that in the first, second, and third terms of such a series, on the assumption that H is always monad and C tetrad, there cannot be any other arrangement of the symbols. In other words, there can only be one compound of the formula CH_4, one C_2H_6, and one C_3H_8. Isomerism is impossible. But on arriving at the fourth member of the series, C_4H_{10}, we find it possible to arrange the symbols in a second order as follows :—

$$CH_3-CH\begin{smallmatrix}\diagup CH_3\\ \diagdown CH_3\end{smallmatrix}$$

The fifth term admits of a still larger number of varieties, which may be expressed in a similar manner.

1. $CH_3-CH_2-CH_2-CH_2-CH_3$

2. $CH_3-CH_2-CH\begin{smallmatrix}\diagup CH_3\\ \diagdown CH_3\end{smallmatrix}$

3. $CH_3-C\begin{smallmatrix}\diagup CH_3\\ -CH_3\\ \diagdown CH_3\end{smallmatrix}$

As we advance in the series the number of isomeric hydrocarbons theoretically possible increases very rapidly, and it has been calculated that at the tenth term, $C_{10}H_{22}$, there may be 75 modifications, whilst at the thirteenth, $C_{13}H_{28}$, there may be no fewer than 799 varieties of formula. The number of these hydrocarbons actually known is at present very small.

In the formula for the olefines and for the acetylene series we must admit the supposition that the carbon atoms are linked together by more than one unit of valency.

Ethylene   Acetylene
$CH_2{=}CH_2$ $CH{\equiv}CH$

The terpenes are liquid hydrocarbons found in turpentine and in various natural essential oils. They all have the formula $C_{10}H_{16}$, and no homologues have yet been discovered. Their constitution is at present unknown.

On reviewing the foregoing formulæ it will at once be perceived that whilst the paraffins are saturated,[1] and hence are capable of producing only substitution compounds, the remaining series referred to may enter into direct com-

[1] The paraffins are more completely saturated than any other known form of matter, for not only do they resist the attack of ordinary chemical agents but they form no 'molecular' combinations, and even as solvents the liquid members of the series have but limited application.

bination with hydrogen, with the halogens, and other substances. Thus the following formulæ represent saturated compounds :—

Olefines $\quad C_nH_{2n}Br_2$
Acetylene series $\quad C_nH_{2n-2}Br_4.$

The formula for benzene may be represented as follows :—

$$\begin{array}{c} HC—CH \\ \diagup\quad\diagdown \\ HC\quad\quad CH \\ \diagdown\quad\diagup \\ HC=CH \end{array}$$

This formula accounts for the following facts. (1) Benzene is the first term of the series, there being no hydrocarbon homologous with benzene containing fewer than six atoms of carbon. (2) Benzene is capable of combining with six atoms of bromine and no more. (3) Benzene gives rise to a large number of substitution derivatives, but of those in which one hydrogen atom is replaced by one monad atom or group there can be no isomerides, because all the six atoms of hydrogen are similarly situated. Of those substitution derivatives in which two atoms of hydrogen are replaced by two monads there are three isomeric modifications. To explain this we number the points of the hexagon at which the carbon symbols are placed, and omitting for the sake of clearness all but the symbols of the replacing radicles, we see at once in what manner these three modifications of the di-derivatives are related to the original hydrocarbon and to one another.

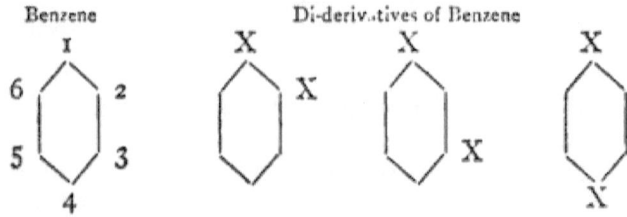

Benzene   Di-derivatives of Benzene

The symbol X represents Cl, Br, I, NO_2, NH_2, CH_3, or other monad radicle. In the first of these formulæ the replacing symbols are represented in the positions 1 and 2, in the second 1 and 3, and in the third 1 and 4. It must, however, be observed that the symbols of carbon to which they are attached are all of the same kind, and it therefore matters not which of these carbons is used as the starting point and in which direction the numbering is carried. The first formula, therefore, merely indicates that the replacing radicles are combined with two contiguous atoms of hydrogen, whilst the second and third formulæ represent them as more widely separated.

In each case the same structure is implied so long as the relative distances from each other of the symbols of X are preserved upon the chain of carbon.

The modifications of the tri- and tetra-substitution derivatives of benzene are accounted for by the same hypothesis.

The higher homologues of benzene result from the introduction of the methyl group, CH_3, in place of one or more atoms of hydrogen. Monomethyl benzene, commonly called toluene, $C_6H_5(CH_3)$, cannot exist in more than one form, as already explained. But the next term of the series, xylene, may exhibit no fewer than four isomeric modifications, as follows :—

Ethyl-benzene $\quad C_6H_5 - CH_2 - CH_3$
Three dimethyl-benzenes $\quad C_6H_4(CH_3)_2$

	Position of methyl
First modification	1 : 2
Second ,,	1 : 3
Third ,,	1 : 4

The number of modifications is still greater among the higher members of the series.

The constitution of the hydrocarbons naphthalene, $C_{10}H_8$, and anthracene, $C_{14}H_{10}$, and their homologues is regarded as closely related to that of benzene, naphthalene being represented as composed of a pair of benzene molecules possessing two atoms of carbon in common:—

$$\text{naphthalene structure}$$

In anthracene two molecules of benzene are connected together by a link composed of two additional atoms of carbon, thus:—

$$\text{anthracene structure}$$

These formulæ indicate that the number of isomeric derivatives of naphthalene and of anthracene must be much larger than in the case of benzene. Many of these have already been isolated and their properties determined. They include a number of important dye-stuffs.

Hydrocarbons differ very much in their physical properties, and in the nature of the reactions to which they lend themselves. This, of course, depends upon the series to which they belong; but taking any one series, such as the paraffins, it is found that a regular gradation may be traced in the physical properties of the several members of the series, passing from the lowest to the highest terms.

Thus the first three members of the paraffin series are gaseous; the succeeding members are liquid, but become less and less volatile,[1] and at the same time more dense and viscid, till among the highest members of the series we

---

[1] See Chapter V., p. 35.

come to crystalline solid bodies, which are volatilisable only at such high temperatures that they cannot be distilled without extensive decomposition. In proportion as we ascend the series the percentage of hydrogen rapidly decreases, and this, together with the diminished mobility of the body, may perhaps account for the comparative indifference to chemical reagents exhibited by the higher members.

A corresponding seriation of properties, both physical and chemical, is observed in other groups of bodies related to one another in a similar manner.

2. ALCOHOLS.

These bodies may be regarded as hydroxides of hydrocarbon radicles, considering that in their most prominent and characteristic reactions they resemble water and the hydroxides of the metals. They are either liquids or solids, and they are obtained in a variety of processes; but the ethylic or common alcohol of fermentation was the first studied, and is the best known of the class. Taking it as the representative of alcohols in general, we may record the following as its characteristic reactions :—

(a) By the action of an alkali metal it loses hydrogen.

$$C_2H_5HO + K = C_2H_5KO + H.$$
Alcohol. Potassic ethylate.

(b) By the action of acids or acid anhydrides it is converted into salts, called compound ethers, e.g.:

$$C_2H_5HO + HCl = C_2H_5Cl + H_2O.$$
Alcohol. Ethyl chloride.

$$C_2H_5HO + H_2SO_4 = C_2H_5HSO_4 + H_2O.$$
Alcohol. Ethyl-hydrogen sulphate.

$$2C_2H_5HO + (C_2H_3O)_2O = 2C_2H_5C_2H_3O_2 + H_2O.$$
Alcohol. Acetic anhydride. Ethyl acetate.

(c) The alcohol is reproduced from these salts by the action of alkalies :

$$C_2H_5Cl + KHO = C_2H_5HO + KCl.$$

Alcohols are divisible into several classes, according to the number of hydroxyl groups in the molecule. Monohydric alcohols yield one saline derivative or compound ether, dihydric alcohols give two, trihydric alcohols give three such compounds, and so on.

The following formulæ, which represent three typical alcohols and the chlorides derivable from them, will serve to illustrate this statement.

$$
\begin{array}{ccc}
\text{Common alcohol.} & \text{Glycol.} & \text{Glycerol.} \\
\text{Monohydric.} & \text{Dihydric.} & \text{Trihydric.} \\
C_2H_5HO & C_2H_4\begin{cases} HO \\ HO \end{cases} & C_3H_5\begin{cases} HO \\ HO \\ HO \end{cases} \\
& & C_3H_5\begin{cases} HO \\ HO \\ Cl \end{cases} \\
& C_2H_4\begin{cases} HO \\ Cl \end{cases} & C_3H_5\begin{cases} HO \\ Cl \\ Cl \end{cases} \\
C_2H_5Cl & C_2H_4\begin{cases} Cl \\ Cl \end{cases} & C_3H_5\begin{cases} Cl \\ Cl \\ Cl \end{cases}
\end{array}
$$

But alcohols are also divisible into several other groups, which are characterised by their behaviour when submitted to oxidation.

Primary alcohols, when oxidised, lose hydrogen, and yield bodies which are called aldehyds ; thus common alcohol loses two atoms of hydrogen :

$$\underset{\text{Alcohol.}}{C_2H_6O} + O = \underset{\text{Aldehyd.}}{C_2H_4O} + OH_2.$$

Secondary alcohols, under the influence of similar reagents, yield ketones ; for example :

Primary, Secondary, and Tertiary Alcohols.

$$C_3H_8O + O = C_3H_6O + OH_2.$$
Isopropylic alcohol. Acetone.

Tertiary alcohols, when oxidised, break up into a mixture of acids, each containing a smaller number of carbon atoms than the alcohol.

These differences are accounted for by the following hypothesis. The generating hydrocarbon is supposed to be marsh-gas or methane, and by the replacement of one of its hydrogen atoms by hydroxyl, an alcohol called methylic alcohol is formed.

$$\text{Methane} \qquad\qquad \text{Methylic hydroxide or methylic alcohol.}$$
$$C \begin{cases} H \\ H \\ H \\ H \end{cases} \qquad\qquad C \begin{cases} H \\ H \\ H \\ OH \end{cases}$$

By replacing another atom of hydrogen by a hydrocarbon group, one of the higher homologues of methylic alcohol is produced. Now, so long as the replacement is effected in such a manner as to leave two of the hydrogen atoms of the marsh-gas type undisturbed, the product is a primary alcohol, which by the substitution of an atom of oxygen for these two atoms of hydrogen is capable of giving rise to an aldehyd.

$$\text{Primary alcohols.} \qquad\qquad\qquad \text{Aldehyds.}$$
$$C \begin{cases} CH_3 \\ H \\ H \\ OH \end{cases} \quad \begin{cases} C(CH_3)H_2 \\ H \\ H \\ OH \end{cases} \text{\&c.} \quad C \begin{cases} CH_3 \\ O \\ H \end{cases} \quad C \begin{cases} C(CH_3)H_2 \\ O \\ H \end{cases} \text{\&c.}$$
Ethylic. Propylic. Acetic. Propionic.

If, however, the second step in the process of replacement affects one of these typical hydrogen atoms, then a secondary alcohol is produced, and this body cannot take in oxygen in the same manner as the primary alcohols, but it may lose two atoms of hydrogen, as represented by the following formula:

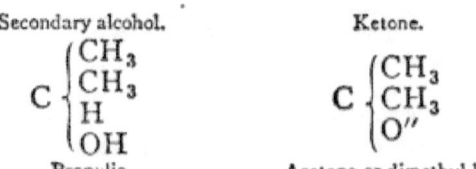

Secondary alcohol. — Propylic.
Ketone. — Acetone or dimethyl-ketone.

Lastly, the tertiary alcohols are generated in the same manner by the replacement of the last atom of hydrogen originally belonging to the type, thus:

Tertiary alcohols.

$$C \begin{cases} CH_3 \\ CH_3 \\ CH_3 \\ OH \end{cases} \qquad C \begin{cases} C(CH_3)H_2 \\ CH_3 \\ CH_3 \\ OH \end{cases} \&c.$$

Butylic.   Amylic.

It is obvious from this that, although a primary alcohol may contain one or two atoms of carbon, no secondary alcohol containing less than three atoms, and no tertiary alcohol containing less than four atoms of carbon in the molecule can exist.

Another class of alcohols, called phenols, is also known. These agree to some extent with secondary alcohols, inasmuch as they yield no aldehyd by oxidation, but furnish compounds containing the same number of carbon atoms as themselves. These substances, however, are not ketones. They are called quinones.

No phenol is known to contain fewer than six carbon atoms in the molecule. They are regarded as derived from benzene or some analogous hydrocarbon. The best known phenol is the body which usually goes by that name among chemists, or, in common parlance, carbolic acid, C_6H_5OH. In accordance with the formula, it might also be called phenyl hydroxide, or phenyl alcohol. It differs from ordinary alcohols in yielding a great number of substitution derivatives, most of which exhibit well-marked acid properties. Thus the trinitrophenol is commonly called picric, or trinitrophenic acid. The following are the formulæ of its chloro- and nitro derivatives:

Monochlorophenol . $C_6H_4Cl.OH$
Dichlorophenol . $C_6H_3Cl_2.OH$
Trichlorophenol . $C_6H_2Cl_3.OH$
Pentachlorophenol . $C_6Cl_5.OH$
Mononitrophenol . $C_6H_4(NO_2).OH$
Dinitrophenol . $C_6H_3(NO_2)_2.OH$
Trinitrophenol . $C_6H_2(NO_2)_3.OH$

Several isomeric modifications of the mono- and di-derivatives are known.

3. ETHERS.

These bodies have the same relation to alcohols as the metallic oxides to their hydrates. They are, however, not easily converted into alcohols by the direct action of water, neither are they obtained from alcohols so readily as the metallic oxides from hydrates. The oxides corresponding with the dihydric alcohols seem to undergo this transformation more readily than those belonging to the monohydric alcohols. The following are some examples of simple and mixed ethers:

$$\left.\begin{array}{l}CH_3\\CH_3\end{array}\right\}O \qquad \left.\begin{array}{l}CH_3\\C_2H_5\end{array}\right\}O \qquad \left.\begin{array}{l}C_2H_5\\C_2H_5\end{array}\right\}O$$

Methylic oxide or ether.   Methyl-ethylic oxide or ether.   Ethylic oxide (common ether).

$$(C_2H_4)''O \qquad\qquad (C_3H_5)_2''O_3$$

Ethylene oxide or ether.   Glyceryl oxide or ether.

The known ethers are volatile liquids of aromatic odour.

4. ALDEHYDS AND KETONES.

These two classes of compounds are closely related, both as regards their mode of formation, and properties and composition.

Aldehyds are formed by the oxidation of primary alcohols, ketones by the oxidation of secondary alcohols. Thus common aldehyd results from the removal of two atoms of hydrogen from common (ethylic) alcohol :—

$$C_2H_6O \;-\; H_2 \;=\; C_2H_4O$$

Acetone, the best known of the ketones, stands in the same relation towards isopropylic alcohol :—

$$C_3H_8O - H_2 = C_3H_6O$$

It has already been shown that no secondary alcohol can contain less than three atoms of carbon, and consequently there can be no ketone, properly so called, containing less carbon than acetone, C_3H_6O.

The formation of this body by the action of zinc methyl (see *organo-metallic compounds*) on carbonyl chloride shows that it contains two methyl groups united with the carbonyl radicle :—

$$CO\begin{Bmatrix}Cl\\Cl\end{Bmatrix} + Zn\begin{Bmatrix}CH_3\\CH_3\end{Bmatrix} = ZnCl_2 + CO\begin{Bmatrix}CH_3\\CH_3\end{Bmatrix}$$

Carbonyl chloride.   Zinc methyl.   Zinc chloride.   Acetone.

It thus appears that the aldehyds of the monocarbon and dicarbon series may be considered to be homologous with the higher aldehyds on the one hand, and with the ketones on the other. The following formulæ indicate the relationship :—

Formic aldehyd or ketone. $CO\begin{Bmatrix}H\\H\end{Bmatrix}$ (gaseous)

Acetic aldehyd or ketone. $CO\begin{Bmatrix}H\\CH_3\end{Bmatrix}$ (B.P. 22°)

Aldehyds.	Ketones.
Propionic. $CO\begin{Bmatrix}H\\C_2H_5\end{Bmatrix}$ (B.P. 48°·5)	Dimethyl (Acetone). $CO\begin{Bmatrix}CH_3\\CH_3\end{Bmatrix}$ (B.P. 56°)
Butyric. $CO\begin{Bmatrix}H\\C_3H_7\end{Bmatrix}$ (B.P. 75°)	Methyl-ethyl. $CO\begin{Bmatrix}CH_3\\C_2H_5\end{Bmatrix}$ (B.P. 81°)
Valeric. $CO\begin{Bmatrix}H\\C_4H_9\end{Bmatrix}$ (B.P. 102°)	Metameric $\begin{cases} \text{Diethyl.} \\ CO\begin{Bmatrix}C_2H_5\\C_2H_5\end{Bmatrix}\text{(B.P. 100°)} \\ \text{Methyl-propyl.} \\ CO\begin{Bmatrix}CH_3\\C_3H_7\end{Bmatrix}\text{(B.P. 101°)} \\ \text{Methyl-isopropyl.} \\ CO\begin{Bmatrix}CH_3\\CH(CH_3)_2\end{Bmatrix}\text{(B.P. 93°)} \end{cases}$

From the existence of such intimate relations between these two classes of compounds it is not extraordinary that in many of their properties there should be considerable resemblance between them.

(a) By the action of nascent hydrogen, generated by the action of sodium amalgam in the presence of water, aldehyds and ketones are converted into the corresponding alcohols, e.g. :—

$$C_2H_4O + H_2 = C_2H_6O$$
Acetic aldehyd. Ethyl alcohol.

$$C_3H_6O + H_2 = C_3H_8O$$
Acetone. Isopropyl alcohol.

(b) All aldehyds and many ketones combine with the acid sulphite of sodium, forming crystallisable compounds, which yield up the aldehyd or ketone again by treatment with an acid or an alkaline carbonate.

(c) By boiling with hydrocyanic acid and hydrochloric acid, aldehyds and ketones are converted into acids containing in the molecule one more atom of carbon than themselves. For example, common aldehyd is converted into one form of lactic acid :

$$CO\begin{cases}CH_3\\H\end{cases} + HCN + 2H_2O + HCl$$
Aldehyd. Hydrocyanic acid.

$$= \begin{cases}CH_3\\CH\end{cases}\begin{matrix}OH\\CO.OH\end{matrix} + NH_4Cl$$
Ethylidene lactic acid. Ammonium chloride.

Acetone is converted by a similar process into a homologue of lactic acid :

$$CO\begin{cases}CH_3\\CH_3\end{cases} + HCN + 2H_2O + HCl$$
Acetone. Hydrocyanic acid.

$$= \begin{cases}CH_3\\C\end{cases}\begin{matrix}CH_3\\OH\\CO.OH\end{matrix} + NH_4Cl$$
Acetonic or oxybutyric acid. Ammonium chloride.

Notwithstanding these various points of resemblance, ketones are sharply distinguished from aldehyds by the action of agents of oxidation.

An aldehyd is always converted by oxidation into an acid containing the same number of atoms of carbon.

A ketone, on the contrary, is much more difficult to oxidise, and is then broken up into a mixture of acids, each containing a smaller number of carbon atoms. This oxidation proceeds according to a definite law.

5. ACIDS.

Every carbon acid may be considered to have been derived from an alcohol by the substitution of an atom of oxygen for two atoms of hydrogen.

Monohydric alcohols yield monobasic acids. For example, ethylic alcohol yields acetic acid:

$$\underset{\text{Ethyl alcohol.}}{C_2H_5.OH} - H_2 + O = \underset{\text{Acetic acid.}}{C_2H_3O.OH}$$

Dihydric alcohols may yield either monobasic or dibasic acids. Ethylene alcohol or glycol, for example, may be converted into monobasic glycolic or dibasic oxalic acid, thus:

$$\underset{\text{Ethylene alcohol.}}{C_2H_4(OH)_2} - H_2 + O = \underset{\text{Glycolic acid.}}{C_2H_2O(OH)_2}$$

$$\underset{\text{Ethylene alcohol.}}{C_2H_4(OH)_2} - 2H_2 + O_2 = \underset{\text{Oxalic acid.}}{C_2O_2(HO)_2}$$

Trihydric alcohols may yield theoretically either monobasic, dibasic, or tribasic acids. Very few of these derivatives, however, are actually known. Glycerol, a trihydric alcohol, yields, upon oxidation, monobasic glyceric acid and dibasic **tartronic** acid:

$$\underset{\text{Glycerol.}}{\begin{array}{c}CH_2.OH\\|\\CH.OH\\|\\CH_2.OH\end{array}} \quad \underset{\text{Glyceric acid.}}{\begin{array}{c}CH_2.OH\\|\\CH.OH\\|\\CO.OH\end{array}} \quad \underset{\text{Tartronic acid.}}{\begin{array}{c}CO.OH\\|\\CH.OH\\|\\CO.OH\end{array}}$$

A large number of acids are known, the alcohols corresponding to which have not yet been discovered.

Acids, however, may be generated by an entirely different process from alcohols containing one atom less carbon. This process is interesting, because it serves to indicate the general constitution of carbon acids.

Starting from methyl alcohol, $CH_3.OH$, for example, we may convert this body into the corresponding cyanide, $CH_3.CN$. When boiled with caustic potash this compound undergoes decomposition, ammonia is evolved, and the potassium salt of acetic acid is generated:

$$CH_3.CN + KHO + H_2O = CH_3.CO(OK) + NH_3$$
Methyl cyanide.   Potash.   Water.   Potassium acetate.   Ammonia.

Or, leaving the potassium out of sight, the reaction may be represented as follows:—

$$CH_3.CN + 2H_2O = CH_3.CO(OH) + NH_3$$

From this it appears that the acid is generated by the substitution of $O'' + (OH)'$ for the N''' of the cyanide. Acetic acid, then, consists of a methyl group, CH_3, combined with a *carboxyl* group, $CO.OH$. Now methylic alcohol is monohydric, and therefore is capable of yielding one cyanide and no more. Consequently, from methylic alcohol only one derivative containing the carboxyl group can be formed, and that derivative is a monobasic acid. In a similar manner it is found that, by decomposing the cyanide of a bivalent alcohol radicle by boiling alkali, a bibasic acid is produced, whilst the cyanide of a trivalent radicle yields by the same process a tribasic acid. Setting down examples of compounds generated in this way, we find that they all contain the group $CO.OH$, once, twice, or three times in the molecule, according to the saturating power of the acid.

Monobasic acetic acid.	Dibasic succinic acid.[1]	Tribasic tricarballylic acid.
$CH_3 \cdot CO(OH)$	$C_2H_4 \begin{cases} CO(OH) \\ CO(OH) \end{cases}$	$C_3H_5 \begin{cases} CO(OH) \\ CO(OH) \\ CO(OH) \end{cases}$

This carboxyl group, then, is contained in a very large number of carbon acids, and its presence seems to be intimately connected with the development of the acid character in these compounds. We have seen (Phenols, p. 324) that the halogens and the nitroxyl group (NO_2)′, when substituted for hydrogen in certain oxidised bodies, also constitute acidifying agents, but the presence of these radicles in varying quantities does not affect the basicity of the resulting acid; whereas it seems to have been established that an acid which contains the carboxyl, $CO(OH)$, group n times is n basic.

Carbon acids, when submitted to the action of the halogens, yield a great many substitution derivatives, such as the following chlorinated compounds, which are obtained from acetic acid:—

Monochloracetic acid	$CH_2Cl \cdot COOH$
Dichloracetic acid	$CHCl_2 \cdot COOH$
Trichloracetic acid	$CCl_3 \cdot COOH$

But here the process of substitution stops, the hydrogen of the carboxyl being replaceable only by metallic or positive radicles, and not by negative bodies.

The metallic salts, amides, anhydrides, and chlorides derived from a carbon acid, correspond in number, mode of formation, and general properties with the same derivatives of mineral acids.[2]

6. BASIC DERIVATIVES OF AMMONIA.

The constitution and general properties of these bodies have already been discussed (Amines, p. 309).

---

[1] See Molecular Weights, Chap. XVI., p. 129.
[2] Pp. 296 to 303.

The 'alkaloids,' quinine, morphine, strychnine, and many others which are obtained from various plants, and constitute important medicinal agents, are referable to this type. They are for the most part tertiary amines.

7. COMPOUND ETHERS OR ETHEREAL SALTS.

The most general method for the formation of these compounds consists in heating together an acid and an alcohol. A double decomposition ensues; water is formed, together with the ethereal salt; so that the reaction is exactly parallel with the change which occurs when an ordinary acid and basic oxide are brought into contact. The following equations, for example, are strictly comparable:—

$$KHO + HNO_3 = KNO_3 + H_2O$$
Basic oxide.   Acid.   Salt.   Water.

$$C_2H_5.HO + HC_2H_3O_2 = C_2H_5.C_2H_3O_2 + H_2O$$
Alcohol.   Acid.   Ethereal salt.   Water.

Compound ethers also imitate metallic salts in their general reactions. Thus they are decomposed by acids, by alkalies, and by water in precisely the same manner. They differ from metallic salts, however, in many physical characters, being for the most part volatile liquids of aromatic odour, and very slightly soluble in water, though generally miscible with alcohol in all proportions. Comparatively few are crystallisable. Among the most interesting examples of crystallisable ethereal salts are the constituents of the natural fats:—

Palmitin
or glyceric palmitate
$$\begin{matrix} C_{15}H_{31}.CO.O \\ C_{15}H_{31}.CO.O {-} (C_3H_5)''' \\ C_{15}H_{31}.CO.O \end{matrix}$$

and

Stearin or
glyceric stearate
$$\begin{matrix} C_{17}H_{35}.CO.O \\ C_{17}H_{35}.CO.O {-} (C_3H_5)''' \\ C_{17}H_{35}.CO.O \end{matrix}$$

8. ORGANO-METALLIC COMPOUNDS.

These are compounds of metals or metalloids with hydrocarbon radicles. They are volatile, very oxidisable, and generally heavy liquids. The zinc-ethyl compound was first obtained. It is a colourless liquid, boiling at 118°, and spontaneously inflammable when thrown into the air. It is instantly decomposed by water, and when submitted to the action of successive small quantities of oxygen it yields two oxidised compounds, thus:—

$$\text{Zinc ethide.} \quad \text{Zinc ethyl-ethylate.} \quad \text{Zinc ethylate.}$$
$$\text{Zn}\begin{cases}C_2H_5\\C_2H_5\end{cases} \quad \text{Zn}\begin{cases}C_2H_5\\OC_2H_5\end{cases} \quad \text{Zn}\begin{cases}OC_2H_5\\OC_2H_5\end{cases}$$

Iodine removes from it half and then the whole of the ethyl, according to these equations:—

$$\text{Zn}(C_2H_5)_2 + I_2 = \underset{\text{Zinc ethiodide.}}{\text{Zn}(C_2H_5)I} + \underset{\text{Ethyl iodide.}}{C_2H_5I}$$

$$\text{Zn}(C_2H_5)I + I_2 = \underset{\text{Zinc iodide.}}{\text{ZnI}_2} + \underset{\text{Ethyl iodide.}}{C_2H_5I}$$

The existence of these compounds furnishes conclusive testimony of the diadic character of the zinc atom, and in other cases similar evidence, of great value in settling questions relating to valency and atomic weight, has been obtained by the study of these compounds.

The following are the formulæ of some of the most interesting organo-metallic bodies. It will be noticed that they are constituted in the same way as the chlorides and oxides of the same hydrocarbon radicles, the special peculiarities of the organo-metallic bodies being due to the unoxidised condition of the metals they contain.

The vapour densities of all these compounds are normal, so that each formula represents two volumes of vapour.

Ethyl chloride
or chlorine ethide . . $\overset{'}{\text{Cl}}-C_2H_5$

Ethyl oxide . .
or oxygen ethide . . $\overset{''}{\text{O}}\begin{cases}C_2H_5\\C_2H_5\end{cases}$

Organo-metallic Compounds. 333

Boron ethide . . . $\overset{\prime\prime\prime}{\text{B}}\begin{cases}C_2H_5\\C_2H_5\\C_2H_5\end{cases}$

Silicon ethide . . . $\overset{\text{iv}}{\text{Si}}\begin{cases}C_2H_5\\C_2H_5\\C_2H_5\\C_2H_5\end{cases}$

Trimethyl arsine
or arsenious trimethide . $\overset{\prime\prime\prime}{\text{As}}\begin{cases}CH_3\\CH_3\\CH_3\end{cases}$

Diarsenious tetramethide .
(kakodyl) $\begin{cases}\overset{\prime\prime\prime}{\text{As}}(CH_3)_2\\\overset{\prime\prime\prime}{\text{As}}(CH_3)_2\end{cases}$

Trimethyl stibine
or antimonious trimethide $\overset{\prime\prime\prime}{\text{Sb}}\begin{cases}CH_3\\CH_3\\CH_3\end{cases}$

Triethyl bismuthine
or bismuthous triethide . $\overset{\prime\prime\prime}{\text{Bi}}\begin{cases}C_2H_5\\C_2H_5\\C_2H_5\end{cases}$

Zinc ethide . . . $\text{Zn}\begin{cases}C_2H_5\\C_2H_5\end{cases}$

Mercuric ethide . . $\text{Hg}\begin{cases}C_2H_5\\C_2H_5\end{cases}$

Stannous ethide . . $\overset{\prime\prime}{\text{Sn}}\begin{cases}C_2H_5\\C_2H_5\end{cases}$

Stannic ethide . . $\overset{\text{iv}}{\text{Sn}}\begin{cases}C_2H_5\\C_2H_5\\C_2H_5\\C_2H_5\end{cases}$

Plumbic ethide . . $\overset{\text{iv}}{\text{Pb}}\begin{cases}C_2H_5\\C_2H_5\\C_2H_5\\C_2H_5\end{cases}$

EXERCISES ON SECTION VI.

1. Write down the formulæ of all the possible silver salts of the following acids :—

$HClO_3$, $HClO_4$, H_2SO_4, H_3PO_4, H_4SiO_4, H_2SO_3, H_3PO_3.

2. Write down the formulæ for all the chlorides and amides theoretically derivable from the following acids :—

HNO_3, H_2SO_4, $H_4P_2O_7$, $HC_7H_5O_2$ (benzoic acid, monobasic).

3. Write the constitutional formulæ of Nordhausen sulphuric acid, $H_2S_2O_7$, of permanganic acid, $H_2Mn_2O_8$, and of chromic acid, H_2CrO_4, assuming sulphur, manganese, and chromium hexad.

4. Write down the formulæ for the possible anhydrides derived from orthosulphuric acid, $S(OH)_6$, by successive stages of dehydration.

5. Give the formulæ for the anhydrides corresponding with sulphuric acid, H_2SO_4, phosphoric acid, H_3PO_4, pyrophosphoric acid, $H_4P_2O_7$, nitric acid, HNO_3, nitrous acid, HNO_2, hyponitrous acid, HNO, theiosulphuric acid, $H_2S_2O_3$, chloric acid, $HClO_3$, perchloric acid, $HClO_4$, and acetic acid, $HC_2H_3O_2$, pointing out those which are actually known.

6. Give concise definitions of the terms *acid*, *base*, *salt*, with examples of their characteristic properties drawn from your own experience.

7. How would you explain the terms acid, and neutral salt? Give examples.

8. A number of salts having the general formula $M'HSO_2$ were called by the discoverer hydrosulphites. What name should they bear according to the rules laid down in Chap. XIII.? Write the graphic formulæ for these bodies, and mention compounds which are similarly constituted.

9. Complete the following equations, adding the name to each formula :—

$HNO_3 + PbO =$
$HNO_3 + NH_3 =$
$H_2SO_4 + KHO =$
$H_2SO_4 + PCl_5 =$
$(NH_4)_2SO_4 - 2H_2O =$
$(NH_4)HSO_4 - H_2O =$
$(NH_4)_2CO_3 - H_2O =$
$(NH_4)_2CO_3 - 2H_2O =$
$C_2H_3OCl + OH_2 =$
$Ca(HO)_2 - H_2O =$

Exercises on Section VI.

$$CaCl_2 + Na_2CO_3 =$$
$$CaCO_3 + 2NH_4Cl =$$
$$NH_3 + C_2H_5I =$$
$$NH_2C_2H_5 + C_2H_5I =$$
$$NH(C_2H_5)_2 + C_2H_5I =$$
$$N(C_2H_5)_3 + C_2H_5I =$$
$$N(C_2H_5)_4I + AgHO =$$

10. Give reasons in favour of each of these formulæ for sal-ammoniac :—

$$NH_3.HCl$$
$$NH_4Cl$$

11. To which class of compounds would you refer the following substances ?—

Caustic potash, KHO ; lime, CaO ; nitrous oxide, N_2O ; nitrogen trioxide, N_2O_3 ; ferric oxide, Fe_2O_3 ; chromium trioxide, CrO_3 ; wood spirit, $CH_3.OH$; glycerine, $C_3H_5(OH)_3$; urea, $CO(NH_2)_2$.

12. Accepting the formula HCl for hydrochloric acid, what facts determine the choice of the formula $H_2C_2O_4$ for oxalic acid ?

13. How far do the following compounds agree with definitions of the terms acid, base, salt ? Give sufficient reasons in each case.

Chlorides of hydrogen, potassium, aluminium, phosphorus, sulphur ; oxides of hydrogen, potassium, carbon ; hydroxides of potassium, aluminium, arsenic, silicon.

14. Define the terms hydrate and hydroxide. Classify the following compounds :—

$$Cl_2.10H_2O, \quad KHO, \quad H_3AlO_3, \quad C_2H_5OH.$$

15. Write the constitutional formula of urea, and of the compound with which it is isomeric.

16. How many litres of marsh-gas would be equal in weight to 25 litres of ethylene ?

17. Benzoate of silver has the formula $C_7H_5AgO_2$. Calculate its percentage composition, also the weight of silver contained in ·5736 gram of this salt.

18. From the formulæ HBr, H_2S, CCl_4, NH_4Cl, H_2SO_4, state the valence of the symbols Br, S, C, N, and S, giving your reasons in each case.

19. Name and classify the following compounds :—

$$CH_4, \qquad CHCl_3, \qquad C_2H_6$$
$$C_2H_5Cl, \qquad \left.\begin{array}{c}C_2H_5\\C_2H_5\end{array}\right\}O, \qquad \left.\begin{array}{c}C_2H_5\\C_2H_5\end{array}\right\}Zn,$$
$$\left.\begin{array}{c}C_2H_5\\H_2\end{array}\right\}N, \qquad \left.\begin{array}{c}CH_3\\(C_2H_5)_2\end{array}\right\}N, \qquad \left.\begin{array}{c}CH_3\\C_2H_5O\end{array}\right\}O.$$

20. Assign a formula to a body containing C 26·6, H 2·2, and O 71·1 per cent.

21. How would you prove that the formula of marsh-gas is CH_4? Also, how could you show that ethylene contains twice as much carbon?

22. Give reasons for representing oxalic acid as $C_2H_2O_4$, not CHO_2; acetic acid as $C_2H_4O_2$, not CH_2O; ethylene as C_2H_4, not CH_2.

23. By what experiments could you prove that the molecule of benzene is correctly represented by the formula C_6H_6, not C_3H_3?

24. A body having the composition $C_3H_{12}O$ is an alcohol. Express by equations the probable action upon it of acetic acid, oxalic acid, and oxidising agents.

25. The compound C_3H_6O is either an aldehyd or a ketone: how would you recognise its true character? Write out the constitutional formula of the aldehyd and ketone of this composition, and indicate the constitution of any metameric modification that may exist.

26. Dissect the crude formula $C_7H_{14}O_2$, showing the nature of the several metameric compounds which it may represent.

27. In respect of what properties may you regard alcohols as water molecules in which hydrogen has been replaced by hydrocarbon radicles? Illustrate by reference to common alcohol, C_2H_5OH.

28. The ethylic series of monohydric alcohols are represented by the general formula—

$$C_nH_{2n+1}OH.$$

Write out the formula for the corresponding aldehyds, acids, and acetic derivatives (acetates).

29. How many acids could be derived from the alcohol—

$$\begin{cases} CH_2.OH \\ CH_2.OH, \end{cases}$$

and what would be their formulæ?

30. Carbolic acid is sometimes called phenylic alcohol. In what respects do its alcoholic properties differ from those of the alcohols of the ethylic series, $C_nH_{2n+2}O$?

31. Complete these equations:—

$$CH_3.CO(OH) + PCl_5 =$$
$$CH_3.OH + HCl =$$
$$\begin{matrix} CH_3 \\ CH_3 \end{matrix}\bigg\} O + 2HCl =$$
$$2CH_3.CO(OH) + Ag_2O =$$
$$CH_3.OH + \begin{matrix} C_2H_5O \\ HO \end{matrix}\bigg\} SO_2 =$$
$$CH_3.CO(OCH_3) + NaHO =$$
$$\begin{matrix} CH_3.CO \\ CH_3.CO \end{matrix}\bigg\} O + H_2O =$$

Exercises on Section VI. 337

32. In what respect does prussic acid resemble and how does it differ from hydrochloric and hydrobromic acids?

33. ·3339 gram of a compound gave ·7896 gram of CO_2 and ·2924 gram H_2O. The vapour density (air = 1) was found to be 6·59. Calculate a rational formula.

34. ·1483 gram of ethylated acetone gave ·3799 gram of CO_2 and ·1575 gram of water. Its vapour density (air = 1) was 2·951. Calculate the formula of the compound.

35. The analysis of barium diacetotartrate gave the following results:—

I. ·2065 gram of substance gave ·1297 gram of $BaSO_4$.

II. ·1377 gram of substance gave ·1307 of CO_2 and ·0300 of H_2O.

Calculate the formula of the salt and of the corresponding acid.

36. The following results were obtained in the analysis of the copper salt of ethyl-benzoic acid: ·2218 gram gave ·4827 gram of CO_2, ·1004 gram of water, and ·0489 gram of CuO. Calculate the formula of the acid.

INDEX.

ABSOLUTE temperatures, 51
Absorption spectra, 59
Acids, 296-328
Adhesion and chemical affinity, 209
Affinity, 5, 203
Alcohols, 321
Aldehyds, 325
Alkaline earths, 261
Alkalis, 259
Allotropy, 199
Aluminium, 272-273
Amides, 312
Amines, 308
Ammonia, derivatives of, 307-330
Ammonium theory, 307
Analysis and synthesis, 5
Antimony, 250
Arsenic, 250
Arsines, 312
Asymmetric carbon, 197
Atomic theory, 85
— weights, 109
Atomicity, 140
Atoms, 4
Avidity, 213

BARIUM, 261
Bases, 296-303
Basic salts, 305
Basigenic elements, 218
Benzene, constitution of, 318
Bismuth, 250
Boiling points, 34
Boron, 235
Boyle's law, 46
Bromine, 220

CADMIUM, 262
Cæsium, 259
Calcium, 261
Carbon, 236
— compounds, 314
Catalytic actions, 165
Charles's law of expansion of gases, 48

Chemical affinity, 5, 203
— combination, laws of, 79
— compounds, characters of, 92
Chemistry, province of, 5
Chlorine, 220
Chromium, 273
Classification of carbon compounds, 314
— of compounds, 296
— of elements, 217
— of reactions, 89
Cobalt, 273
Coefficient of absorption of gases, 27
Cohesion, 7
Colloids, 30
Combustion, 186
Compound ethers, 331
— radicles, 148
Compounds and mixtures, 92
Conditions of chemical change, 161
Constitution distinguished from composition, 6
Constitutional formulæ, 144, 197, 198
Copper, 273, 279
Critical point, 10, 38
Cryohydrates, 16
Crystallisation from solutions, 25
Crystalloids, 30

DALTON'S atomic theory, 83
Davy on flame, 64
Davy's experiments on chlorine, 71
Dialysis, 31
Diffusion of gases, 40
— of liquids, 29
Dimorphism, 193
Dissociation, 168
Dumas on atomic weights, 85

EBULLITION, 33
Efflorescence, 170
Electricity and chemical action, 183
Electrochemical theories, 205
Electrolysis, 184, 205
Elements and compounds, 70
— basigenic, 218
— non-metallic, 217
— oxygenic, 217
— table of, 76
Energy and chemical change, 6, 176
— of isomerides, 202
Equations, 86
Ethereal salts, 331
Ethers, 325
Eutectic mixtures, 15
Exercises on Section I., 67
— on Section II., 105
— on Section III., 153
— on Section IV., 215
— on Section V., 293
— on Section VI., 334

FLAME, luminosity of 64
— structure of, 65
— temperature of, 188
Fluorine, 220-224
Formula explained, 78
Fractional distillation, 36
Frankland on flame, 65
Fraunhofer's lines, 60
Fusion, 12

GALLIUM, 272
Gases, 7
— diffusion of, 40
— expansion of, by heat, 48
— liquefaction of, 36
— occlusion of, 44
— passage of, through metals and membranes, 44
— pressure of, 46
— solution of, 26

340   *Chemical Philosophy.*

GAY

Gay-Lussac's law of expansion, 48
— — of volumes, 82
Germanium, 248
Gold, 281
Graham's law of diffusion, 41
Graphic formulæ, 144

HALOGENS 220
   Heats of combustion, 189
— of formation of chlorides, &c., 180
— of neutralisation of acids, 178
Homologous series, 35
Hydrocarbons, 315
Hydrogen, 245

INDIUM, 272
   Iodine, 220–222
Iridium, 282
Iron, 273
Isomerism, 192
Isomorphism, 116

KETONES, 325

LAW of Avogadro, 51
   — of Boyle, 46
— of Charles, 48
— of chemical combination, 79
— of diffusion of gases, 41
— of Dulong and Petit, 114
— of even numbers, 146
— of Gay-Lussac (law of volumes), 83
— of isomorphism, 116
— of volumes (Gay-Lussac), 83
Lead, 268
Liquefaction of gases, 38
Liquid diffusion, 29
Liquids, 7
Lithium, 259
Luminosity of flame, 64

MAGNESIUM, 262
   Manganese, 273
Mass, action of, 163
Matter, constitution of, 3
Melting points of mixtures, 14
Mendelejeff's classification of elements, 291
Mercury, 265

MET

Metalloids, 244
Metals, 256
Metamerism, 195
Metric weights and measures, 2
Meyer's curve of atomic weights and volumes, 289
Mixtures and compounds, 98
Molecular compounds, 211, 227
— theory, 8
— weights and formulæ, 122
Molecules, 3
— of elements, 123
Molybdenum, 255

NAMES of compounds, 101
— of elements, 100
Nascent state, 126
Newlands on atomic weights, 288
Nickel, 273
Niobium, 254
Nitrogen, 239
Nomenclature, 99
Non-metals, 219

OCCLUSION of gases, 44
Organo-metallic bodies, 332
Osmium, 282
Oxygen, 227
Ozone, 231

PALLADIUM, 282
   Periodic law, 284
Phlogiston, 187
Phosphines, 312
Phosphorus, 239–241
Physical isomerism, 192
Platinum, 282
Polymerism, 194
Potassium, 259
Prout's hypothesis, 284

RATIONAL formulæ, 137
Reactions classified, 89
Rhodium, 282
Rubidium, 259
Ruthenium, 282

SALTS, 296–305
   Saturating power of acids and bases, 131

ZIR

Selenion, 227
Silicon, 236
Silver, 265
Sodium, 259
Solids and fluids, 6
Solution of gases, 26
— of solids, 16
— theories of, 24
Solutions, supersaturated, 26
Solvents, influence of, 162
Spectra, 53
— absorption, 59
— emission, 53
— invisible, 63
Stibines, 312
Strontium, 261
Substitution compounds, 133
Sulphur, 227–234

TANTALUM, 254
   Tellurium, 247
Thallium, 268
Thermal effects of thermical action, 176
Thermolysis, 169
Tin, 248
Titanium, 248
Tungsten, 254
Types, 136

URANIUM, 255

VALENCY, 139
   — of carbon and silicon, 238
— of common salt radicles, table, 152
— of oxygen and sulphur, 233
— of principal elements, 143
— of the halogens, 225
— of the iron group, 277
— of tin and its allies, 247
Vanadium, 250
Vapour densities, abnormal, 171
— density, 123

WATER of crystallisation, 25
Weights and measures, table of, 2

ZINC, 262
   Zirconium, 248

Spottiswoode & Co. Printers, New-street Square, London.

www.ingramcontent.com/pod-product-compliance
Lightning Source LLC
Chambersburg PA
CBHW020243240426
43672CB00006B/625